KOMBUCHA!

KOMBUCHA!

The Amazing Probiotic Tea That Cleanses, Heals, Energizes, and Detoxifies

ERIC AND JESSICA CHILDS

AVERY
a member of Penguin Group (USA)
New York

Published by the Penguin Group
Penguin Group (USA) LLC
375 Hudson Street
New York, New York 10014, USA

USA • Canada • UK • Ireland • Australia
New Zealand • India • South Africa • China

penguin.com
A Penguin Random House Company

Library of Congress Cataloging-in-Publication Data

Childs, Eric, author.
Kombucha! : the amazing probiotic tea that cleanses, heals, energizes, and detoxifies / Eric and Jessica Childs.
p. cm
Includes index.
ISBN 978-1-58333-531-4
1. Kombucha tea. 2. Tea fungus—Therapeutic use.
I. Childs, Jessica, author. II. Title.
RM666.T25C45 2013 2013020753
615.3'21—dc23

Printed in the United States of America
1 3 5 7 9 10 8 6 4 2

Book design by Meighan Cavanaugh

The recipes contained in this book are to be followed exactly as written. The publisher is not responsible for your specific health or allergy needs that may require medical supervision. The publisher is not responsible for any adverse reactions to the recipes contained in this book.

While the authors have made every effort to provide accurate telephone numbers, Internet addresses, and other contact information at the time of publication, neither the publisher nor the authors assume any responsibility for errors, or for changes that occur after publication. Further, the publisher does not have any control over and does not assume any responsibility for author or third-party websites or their content.

This book is dedicated to Brooklyn,

the borough that nourishes our family

so we can nourish yours.

CONTENTS

FOREWORD

The first time I tried kombucha, I was hooked. Not only is it delicious, but I quickly realized that it had great health and performance benefits for an endurance athlete like me.

We've all heard, "You are what you eat," but perhaps it should be, "You are what you digest." Digestibility is key to high-net-gain nutrition and helps release the largest yield of micronutrients, a concept I discuss further in my book *Thrive*. If your body can't break down your food and assimilate it, the nutritional value cannot be realized. I found that adding kombucha to pre-workout fuel speeds up digestion. Uncluttered digestion is important for everyone, but its value is heightened for athletes when they consume their pre-workout fuel. I became such a fan of kombucha's ability to increase the net gain of food that I used it in the formula for my Vega Pre-Workout Energizer.

Another remarkable effect of kombucha is the drastic reduction in inflammation the kombucha drinker experiences from its alkaline-forming nature.

Inflammation reduction is another essential component to overall health. If you're an athlete, it's exponentially more valuable. Increased flexibility and increased range of motion, both benefits of lowering inflammation, translate into better performance while significantly reducing the risk of muscle strain and tear.

Less inflammation also equals enhanced efficiency. If muscles are able to move freely with little force being exerted, that's energy being conserved with every contraction. The benefit is immediately reaped with improved athletic performance.

For non-athletes, think of all the muscle contractions you make throughout a typical day even by just walking across the room. If you aren't suffering from inflammation, you don't work as hard. If you don't spend energy on movement, you still have it in your fuel tank. So kombucha will boost your energy by way of enhanced muscle efficiency, not the short-term stimulation you get from many of the energy drinks on the market.

Eric and Jessica Childs's breadth of knowledge on the topic of kombucha is staggering, from a historical and scientific standpoint, right up to its creative application for modern-world benefit. I've been a fan of their work ever since I had my first bottle of KBBK 'buch back in 2009. If they have their way, everyone will drop the soft drinks and head to the all-natural energy drink that is kombucha, and I support their efforts wholeheartedly.

Brendan Brazier
July 15, 2013

INTRODUCTION

KOMBUCHA, YOUR NEW HOT ROOMMATE

Kombucha is the superfood of beverages. It is probiotic, like yogurt, and is loaded with antioxidants. It detoxifies the liver and blood, it provides a crash-proof energy boost, it focuses the mind, it settles digestion, and it streamlines a wide variety of inefficiencies in the body. All of this in a delicious beverage without the high amounts of caffeine and sugar that jam-pack the drink shelf.

Whether you are taking your first sip while you read this or you have been drinking kombucha since the sixties, kombucha has transcended time and distance to arrive here. From ancient Asian origins, the live fermented tea known as kombucha has been passed down from our ancestors to bestow its life-supportive properties on modern folks. And we sure need it these days in our hustling bustling world of pollutants, hard work, and lack of sleep. Kombucha is an ancient elixir that is excellent support for health and well-being in our modern world.

From celebrity style pages to the homes of families around the world, you will see kombucha in the hands of people from all walks of life. Artists flock to kombucha to help focus their minds and bodies for their intense creative processes. Athletes prefer kombucha as an all-natural energizer and performance enhancer. The elegantly beautiful keep a bottle nearby to help them stay young and gorgeous. Health experts laud kombucha's natural detoxifying properties, including it in their prescription for optimum health. Whatever your reason for trying kombucha, you will be amazed with how easy it is to make and how effective it is in optimizing your health.

As the owners of Kombucha Brooklyn (KBBK), a business whose primary objective is to nourish our world community with delicious kombucha brews, we get to mingle with all sorts of 'buch devotees. At gatherings that we call 'buch brigades, we open the doors to the KBBK Learning Center so kombuchasseurs can come drink special in-house kombucha craft brews, show off their own home brews, and share the happiness and good health that kombucha brings. Yoginis, robot builders, farmers, doctors, skateboarders, construction workers, foodies, retirees, kids, and even dogs (!) show up to these events out of a passion for the beverage. One thing we have learned is that whether you are a fully invested kombuchasseur whose 'buch is an essential part of everyday life or you just like to grab a bottle every now and again after a hard workout, people who drink kombucha do it because it makes their lives better. Every kombucha drinker has his or her own story to tell. Here are ours.

Eric Childs: The Story of the Kombuchman

Before Eric became the Kombuchman, he was working in the high-stakes art world in the heart of Manhattan. Pressure to be as pristine as the porcelain he was handling combined with Eric's acne vulgaris, those painful bright red bumps that nobody wants around, beat down our future Kombuchman and almost stripped him of his mojo. He tried everything to get rid of it. Morning creams, night creams, eye creams, serums, dermabrasions, masks—you name it, he tried it. Hundreds of dollars and countless disappointments later, nothing had helped.

In addition to acne, Eric also suffered from the annoying digestive ailment known as heartburn. Before a meal, during a meal, or between meals, this plague was with him all day, every day. He tried all sorts of remedies, but nothing could soothe his pain.

Eric was introduced to kombucha by a friend in 2007, just as thousands of others had been through the ages. Eric knew right away that the extraordinary-tasting health drink would change his life, and he started drinking it every day. Three weeks in, the blemishes started fading and were replaced by healthy, clean, clear skin! And that constant pesky heartburn? Well, that's ancient history.

Jessica Childs: A Foodie with a Knack for Science

Jessica Childs's kombucha love story began with her taste for food adventures in foreign lands. Some of the most thrilling dishes in the world come from destinations scattered far and wide around our lovely planet, and at the tippy-top of awesome foodstuffs are the plates sold by street vendors—the moms and pops who set up shop on the sidewalk and cook over a coffee-can stove like it's the final round of *Iron Chef.* The smells! The flavors! From aromatic pho and rebel-rousing bia hơi in Hanoi to the wonderful unidentifiable meat soups and "stinky-toe fruit" salads of San Ignacio to the hákarl and brennivín of Reykjavík, Jessica delighted in it.

Jessica's gut, however, was clearly not delighted with her intercontinental adventures. For years, daily bellyaches that inexplicably came and went were status quo. Doctors couldn't find anything to treat, so she just did the best she could and tried not to aggravate her sensitive belly. Coffee was a real problem, though. If she drank any coffee at all, even just a few sips, she was bowled over in excruciating pain. For years Jessica lived with this undiagnosed bellyache.

Jessica eventually became a microbiologist, and as a food-loving scientist, she thought there was nothing more enchanting than culturing stuff in her kitchen—and then eating it! Kombucha became her go-to ferment because it is incredibly easy to make, it tastes awesome, and it imparts a physical fortitude while enhancing mental elasticity. Then one day Jessica noticed that she hadn't noticed anything at all! No bellyache . . . for days. She'd eat breakfast, then carry on. Eat lunch, and oh, nothing special, nothing to report. She was just going about her days with a carefree and happy belly.

Jessica knew fermented foods were good for digestion but had no idea they could have this kind of impact. She started looking into how these microbial wonders work to create order in a chaotic GI tract. The evidence was compelling. After a few months, she mustered the guts (so to speak) to try a little bit of warm, rich coffee. It was delightfully a non-event, but she still prefers kombucha.

KBBK: Kombucha Done Brooklyn Style!

Our company, Kombucha Brooklyn (KBBK), was conceived in a food hot spot. New York City is recognized as one of the most innovative and creative food cities in the world. So what happens when you drop an ages-old kombucha recipe into the hands of a couple of cosmopolitan foodies from Brooklyn? You let loose a torrent of flavors and uses that forever change the way the world sees this health beverage.

We've come a long way since we first developed our simple recipe of hand-selected organic green, black, and white teas. The foundation of all of our kombucha flavors, this triad of teas forms the perfect base for a delicate-tasting drink with a robust nutritional profile. We now sell our kombucha in ready-to-drink bottles, on tap at locations throughout metropolitan New York City, and through our home brew supplies line available in retail stores internationally and online. If you're a kombucha newbie, try brewing some of our Straight-Up following the recipes given in this book. You'll be a convert faster than you can learn to pronounce the word *kombucha*:

kahm-BOO-cha

We have been drinking kombucha for more than a decade and are thoroughly convinced that it has improved our lives by a substantial margin. So much so that we are writing a book about it from our hearts. From healing our painful and irritating ailments to affecting our lives in the more subtle areas of emotion and energy, and even potentially averting an internal toxic disaster, we are confident we are leading the world toward healthier lives by sharing our knowledge in this book. We hope you'll join us and discover the amazing health and delicious taste of kombucha!

About This Book

Our ceaseless tinkering has shown us the rich possibilities of kombucha. We'll share with you our deep exploration into different aspects of brewing, bottling, and flavoring kombucha that results in a superior kombucha beverage—one that is imbued with the layers of attention we have paid to creating a balanced, unparalleled flavor profile, as well as providing the extraordinary health benefits associated with kombucha throughout history.

In 'Buch Kamps I and II, you will learn basic brewing and bottling techniques that will lead you to a variety of healthful and delicious kombucha brews. Oh, and did we mention how easy it is to brew your own kombucha at home using ordinary kitchen ingredients? If you can make a pitcher of tea, you can brew your own kombucha!

In 'Buch Kamp III, we share our most beloved natural flavor combinations using exotic teas, tropical fruits, garden vegetables, and vibrant herbs and spices. With our knowledge in your quiver, once you have graduated from 'Buch Kamp you will be a 'buch brigadier capable of brewing kombucha at the top of its potential.

'Buch brigadiers can then set about charting new kombucha frontiers! In Chapters 6, 7, and 8, you will find that there really are no limits to what you can do with kombucha. Used as the base for a rockin' kocktail, kombucha is a versatile mixer with endless health-promoting, delicious possibilities. We like to call it reverse toxmosis—detoxifying while you are toxifying.

And the awesome news is that kombucha doesn't quit at the bottom of your glass. Kombucha can be used to enliven appetizers, entrées, and even desserts, adding an array of probiotics, antioxidants, and natural organic compounds that bring an ordinary meal to life (literally!).

Ever heard the saying "What's good for you on the inside is good for you on the outside"? This typically means that if you eat well, your skin, hair, and nails will show it. Fans of kombucha throughout the ages have taken this statement to a whole new level by applying kombucha directly to their exterior for show-stopping, jaw-dropping, eye-poppingly radiant looks!

We're going to show you how to *do* kombucha. You'll discover the history and science behind this ancient elixir. And we'll show you how to harness its power, enjoy its essence, and experience the energy and transformation it brings. From the pitcher to the plate, from the kitchen to the bathroom, from the inside to the outside—the power of kombucha has no limits. Discover it for yourself. Before long, you'll have your own story to tell of how kombucha gave you a tasty beverage, a fancy dessert, brighter skin . . . and a better, more vibrant version of yourself.

WHAT EXACTLY *IS* KOMBUCHA?

Intro to Fermentation

Kombucha is a fermented tea beverage that satisfies two cravings: your craving for yumminess and your craving for optimum health. We see your eyebrows raised. Did you get stuck at the word "fermented"? When people hear the words "fermented beverage," they usually think of one thing: beer. Fermentation is used to make several of our favorite foods, both with alcohol and without. Beer, sauerkraut, cheese, bread, wine, yogurt, and even chocolate are all made using fermentation. To answer the question of what kombucha is, we should start by describing the process of fermentation.

Fermentation has been around since the beginning of time. It is the same process our ecosystem uses to "digest" complex organic matter into simpler molecules and make it ready to enter back into the food chain at ground level. A good parallel is happening right now in your compost pile, where leaves and kitchen scraps are being transformed into garden soil.

When talking about fermenting food, there is undesirable fermentation that occurs when bread sits out too long and becomes moldy. Then there is the desirable fermentation that happens when you deliberately set the right conditions for beneficial yeast and bacteria to transform your foodstuff at the molecular level into something delicious and nutritious. Desirable fermentation of food results in high-quality nutrients, interesting flavors, and an extended shelf life.

Fermentation = transformation at the molecular level by live microbes

Food fermentation begins when a living culture of microorganisms, either one species or many, are introduced to a welcoming environment where food is plentiful. In the case of beer, that environment is a slurry of grains, like barley or wheat, that have been broken down into easily accessible sugars during the malting process. Kombucha, on the other hand, starts from a nutrient solution of sweetened brewed tea that we like to call the *nute*, shorthand for nutrient solution.

The sugars and other compounds in both beer and 'buch nutes are gobbled up by the microorganisms and transformed into all sorts of basic nutrients. One way to think of fermentation is that the microbes are doing some of the work of digesting the food for you. They are breaking up larger molecules into smaller, more accessible ones, unlocking minerals and vitamins from the complex molecular cages they are packaged in and transforming sugars into other useful molecules. The microbes in kombucha transform the sugar and other compounds in the nute into highly accessible polyphenols and other antioxidants, B vitamins, vitamin C, organic enzymes, amino acids, and organic acids.

The microbial culture that works this magic in kombucha is a mixture of bacteria and yeasts that come together in one of the most complex dances the fermentation world has shown us. This culture forms what is called a SCOBY (symbiotic culture of bacteria and yeast) every time a new batch of kombucha is brewed. A SCOBY is a cellulose patty that is created by the microorganisms on the surface of fermenting kombucha and is evidence that they're hard at work.

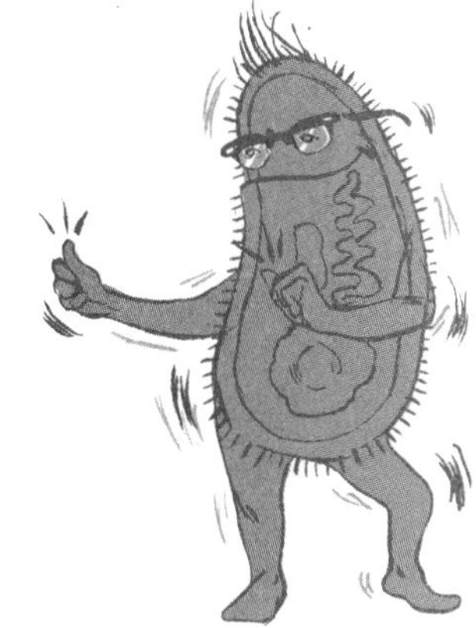

Essentially, Kombucha Is a Superfood

Superfoods are foods dense with nutrients that nourish your body and cause you to feel good. Fresh blueberries, dark chocolate, and kale are all superfoods—kombucha is certainly in good company there. Because kombucha is consumed fresh, when the nutrients have just been made and are still floating in their comfy native environment, protected from contamination and spoilage by the competitive nature of the kombucha microbes, there is no need to diminish the qualities of the brew by pasteurization or further refining. So the nutrients are untouched, unchanged, as perfect as they ever will be, and they are waiting for you to use them!

Processing foods by pasteurizing, homogenizing, refining, or freezing them diminishes their nourishing qualities. High temperatures unravel and otherwise change the composition of enzymes, other proteins, fats, vitamins, and other useful compounds, rendering them less functional or obliterating their usefulness altogether. The difference between a moments-old freshly juiced apple and the pasteurized apple juice you find at the grocery store sitting on the shelves is tremendous. Drink one a day for a week and then try the other the next week. Pay attention to your energy levels, other health indicators that are specific to you (like headaches or muscle aches), and overall mood during the test. It'll change the way you think about eating fresh!

In addition to the aqueous nutrient content, your kombucha will contain the live microorganisms themselves. These tiny little soldiers are of great benefit to your digestive system. Some of the same kombucha microbial species already exist inside your body as part of your gut's microflora. The consumption of live probiotics has been shown to improve how your internal microbial community helps you digest your food. The healing and streamlining effects of probiotics are well known; kombucha is a great source for them.

There are many factors that play into a healthy gut population. Ensuring that you eat foods that have lots of prebiotics (foods that are good nourishment for your internal flora) will also help you achieve a proper balance of internal microflora. Foods that contain good amounts of prebiotics and make great kombucha flavorings are berries, bananas, and cherries.

Kombucha Defined

Kombucha is a nutritionally alive drink that, through the power of fermentation, nourishes your body with compounds that detoxify, energize, support your immune system, streamline your digestive system and your skin, prevent disease, and elevate your mood. Kombucha makes you a healthier you and your day a more awesome day. Once you start, you'll be hooked for both the way it makes you feel and the way it tastes.

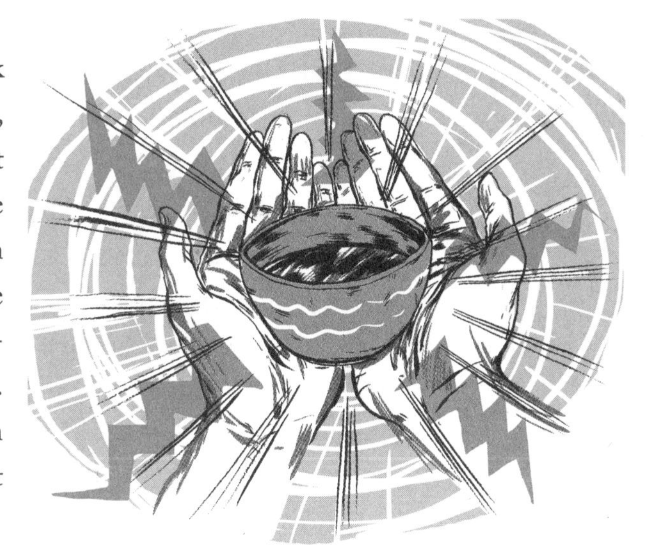

"Kombucha is a tart sunrise bursting on the tongue" . . . "that optimizes the body's performance on the cellular level."

—Eric Childs, interrupted by Jessica Childs

Wow, 'Buch Sounds Awesome! But What's in It?

The legacy of kombucha is a passionate and addicted following of healthy individuals whose beneficial experiences with it has motivated them to share this delicious health brew throughout the ages. Kombucha is not a bottle of aspirin. There is no heavy-handed prescription that says, "Take two and your headache will go away." Kombucha is primarily a food, like yogurt or sauerkraut. As such, its long, successful history of consumption by humans leaves our primary research institutions with no need to delve too deeply into it. However, a search of published scientific papers on kombucha will turn up a respectable showing on the positive effects that kombucha has been shown to have on various diseases. Of particular interest is the substantial and growing body of evidence that kombucha can ameliorate conditions of liver toxicity and exposure to cancer-causing agents. In a world where we are bombarded with carcinogens and toxins from every direction, kombucha looks like it might play the hero's role in preventing some of that damage.

In addition to studying kombucha at the disease level, several research projects have shed light on the diminutive composition of the finished kombucha ferment. Below we have put together a list of components that have been found in reputable studies of kombucha. Not every kombucha brew is the same; in fact, they can differ widely in composition and even in the species of bacteria and yeasts present. The compounds listed below have been found in a high number of brews tested. Every time we read through this list, we're so glad to be kombucha drinkers. And we're glad that we have shared it with the people we love.

After a good chunk of time spent standing over a lab bench observing "life," we've concluded that individual organisms are pretty complex, variable, and unpredictable. Our freaking incredible human bodies are as unique as snowflakes. Find out for yourself what kombucha (and everything else that you eat, drink, breathe, and do) does for you, and find out a little about yourself in the process!

Many claims are made about the presence of particular compounds in kombucha. If the evidence to back up a particular claim was nonexistent, weak, or so obscure that we couldn't get our hands on it, we have not included that claim. Careful attention was paid to include only those compounds that have been verified in reputable studies we could find. That does not mean we do not believe that the other compounds aren't present. It simply means we weren't the ones to put it in writing. If you want to see additional claims, the Internet can help you out with that.

STUDIES SHOW THESE COMPONENTS WILL LIKELY SHOW UP IN YOUR KOMBUCHA

Acetic acid is a powerful antibacterial agent. Acetic acid levels will grow higher as the ferment progresses, making it virtually impossible for non-kombucha microbes to contaminate your brew. It's also responsible for the vinegar-like flavor that kombucha takes on. Studies have shown that acetic acid can have the profound effect of leveling blood sugar spikes by interfering with the breakdown of starches and sugars.

Sour does not mean power! You don't need to ferment your brew until it's vinegar to experience the health benefits of acetic acid. A little bit goes a long way.

Amino acids are the building blocks of proteins. Aside from water, amino acids in the form of proteins make up the largest portion of the human body. In addition to being necessary for building new, useful proteins, amino acids are also key players in a variety of other pathways in the body. From neurotransmission to metabolism, amino acids are critical little molecules.

B vitamins are known to be of such vital importance that the U.S. government has mandated their inclusion in staple products like bread. The family of B vitamins is diverse and affects all sorts of systems in your body. These B vitamins show up in samples of kombucha:

Vitamin B_1, or thiamine, which is used by all living organisms but is made only by plants, bacteria, and fungi. Deficiencies in this nutrient lead to optical, neurological, cardiovascular, and other diseases.

Vitamin B_2, or riboflavin, which has been useful in the treatment of migraine headaches and anemia. It is also an essential element in the breakdown of fats, carbohydrates, and proteins and the production of energy within your body.

Vitamin B_3, or niacin, which significantly decreases the risk of heart disease. That's pretty awesome. Niacin is also helpful in regulating hormones—specifically sex hormones. And you thought it was the alcohol in your kocktail that heightened your mojo!

Vitamin B_5, or pantothenic acid, which is useful in regulating the ratio of good cholesterol to bad cholesterol, thus potentially averting cardiac disaster. One version of the vitamin B_5 molecule, called pantethine, decreases levels of triglycerides in the bloodstream; those nasty buggers increase your risk of heart disease too.

Vitamin B_{12}, or cyanocobalamin, which has the very important job of being integral to proper cell maintenance, particularly in the blood and nervous system. As such, vitamin B_{12} can help in the treatment of fatigue, Alzheimer's disease, breast cancer, and anemia.

Butyric acid is made from glucose by several bacterial strains found in kombucha and also in the human gut. This friendly acid has been shown to nourish healthy gut cells while seeking out and destroying colon cancer cells and inflammation in the same breath.

> **Butyric acid is a great example of what happens to the sugar you add to your kombucha ferment. The sugar is not for you. It's for the microbes to use in making your wonderfully nutritious beverage!**

Caprylic acid, or octanoic acid, has been associated with a reduction in high blood pressure and in the treatment of Crohn's disease. Caprylic acid is also a powerful antimicrobial compound that is used to treat vaginal and other yeast infections like thrush.

Catechins and other polyphenols are found in high concentrations in plants, especially in their leaves. These awesome nutrients prevent oxidation; thus they are antioxidants. Antioxidants have a short life span in your body, so it is good to take them in every few hours to keep up your supply. Another perk of polyphenols? A good amount of evidence shows that catechins and other polyphenols help reduce body fat. Booyah!

Citric acid is naturally found in high concentrations in citrus fruits. Even though it is technically an acid, it functions as an alkalizing agent, restoring balance to overly acidic body fluids such as blood. And as if that wasn't enough, citric acid is a powerful antioxidant, and it also protects the kidneys by binding and removing excess minerals that can build up in them, such as calcium.

Decanoic acid, or capric acid, is found in tropical oils such as superhealthy virgin coconut oil. It has been shown that capric acid helps improve the ratio of "good" (HDL) cholesterol to "bad" (LDL) cholesterol.

Enzymes are protein molecules that are made from amino acids and serve as catalysts for chemical reactions in the body. For pretty much every chemical reaction that takes place in the body at least one enzyme is involved, and in most cases there are many.

Glucaric acid occurs naturally in a variety of fruits, vegetables, and legumes, and your body makes a small amount of it. Glucaric acid has been shown to expedite the detoxification of carcinogens, excess hormones, and other toxins from the liver. It's also currently being investigated for its potential in slowing down or stopping the development of different cancers.

Gluconic acid is a product of the breakdown of glucose by Gluconobacter strains of bacteria that can be found in both your gut and in kombucha. Gluconic acid is thought to interact with butyric acid to improve GI tract health.

Lactic acid is now thought to be purposefully made by muscles during exercise as a highly accessible fuel source to our energy powerhouses, the mitochondria, instead of being a dangerous sign of muscle fatigue, as scientists previously hypothesized. The more fuel mitochondria have, the more work they can do and for longer periods, resulting in improved performance.

We love science. *Love it.* It is incredibly valuable and illuminating. But science is also imperfect and constantly evolving. Throughout the era when scientists thought lactic acid was a sign of muscle fatigue, coaches continued to train athletes in the best way they knew, working on endurance even though they had been told that there was a lactic acid threshold. Like these coaches, one must always pay equal respect to one's own experience and rational judgment alongside research.

Niacinamide is a molecule that is shown to have antianxiety and anti-inflammatory properties, and this little variant is also being looked into for cognition restoration in those suffering from the devastation of Alzheimer's disease.

Oxalic acid has a bad reputation. Although it shows up in many healthy foods such as spinach, apples, raspberries, sorrel, and rhubarb, it is the main building block in kidney stones and a very powerful acid that can be toxic at high levels. As with all things, moderation in life is key. Don't eat too many apples, don't eat too much spinach, and don't drink too much kombucha. You know what's reasonable. Don't overdo stuff. We talk about reasonable amounts of kombucha to drink later in this chapter.

Phenethyl alcohol is a natural compound used as a flavoring agent and a nontoxic ingredient in perfumes because of its sweet floral odor. No wonder kombucha tastes so *gooooood.*

Propanoic acid is a powerful antimold agent that is used commercially to protect foods. Here it occurs naturally and protects your 'buch!

Succinic acid supplementation has been shown to promote neural system recovery and reduce the effects of hangover by accelerating the decomposition of acetaldehyde, a toxic by-product of alcohol metabolism. As if that's not reason enough to want this in your body, succinic acid is a strong antibiotic and an important part of energy production in your body.

A variety of yeast and bacteria species have been found in samples of kombucha. Depending on the heritage of the culture, where and how it was propagated, testing methods, and other factors, scientists have found anywhere between one and four different yeast strains and between two and ten species of bacteria in any given culture. That doesn't mean that more can't be present in a given culture. Specific assays and isolated samples are limited by design. Whether or not we have found the limits of how many species can be present in a single kombucha culture, here are some of the species that have been detected.

Yeasts of the genera *Brettanomyces*, *Zygosaccharomyces*, *Saccharomycodes*, *Saccharomyces*, *Candida*, *Pichia*, and *Schizosaccharomyces* have all been identified in samples of kombucha. Specific species that have been isolated from kombucha samples include *Saccharomyces cerevisiae*, *Pichia fermentans*, *Pichia fluxuum*, *Candida valida*, *Candida kifir*, *Candida lambica*, *Saccharomycodes ludwigii*, *Saccharomycodes apiculatus*, *Schizosaccharomyces pombe*, and *Zygosaccharomyces kombuchaensis*.

The terms *yeast* and *candida* have become buzzwords throughout the health world, signaling a pesky infection that is hard to defeat. The reality is that most yeast species are harmless. Of the varieties of yeast that can be pathogenic, they are called *opportunistic*, meaning that they cause infection in people with compromised immune systems. Most people live quite comfortably with these potentially pathogenic species, even hosting them in low levels as part of the healthy gut flora. For those who experience yeast blooms anywhere in their body, it will be up to you to decide if kombucha is helpful or harmful. Drink, pay attention, and make the decision that works best for you.

Bacteria genera that have been isolated from kombucha samples include *Acetobacter*, *Bacillus*, *Rothia*, and *Gluconacetobacter*. Specific species isolated from within those categories include *Gluconacetobacter xylinus*, *Gluconacetobacter kombuchae*, *Acetobacter xylinoides*, *Acetobacter ketogenum*, *Acetobacter xylinum*, *Bacillus megaterium*, *Bacillus amyloliquefaciens*, and *Rothia dentocariosa*.

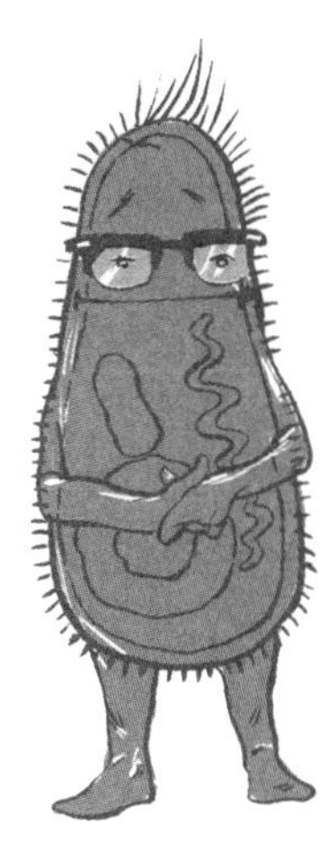

Vitamin C is essential to the production of collagen in your body, and thus is essential to the healing of wounds, inside and out,

and to the maintenance of blood vessels and cartilage. Although most organisms make their own, we are one of the lucky species that don't. Lucky? Well, not if you're a pirate and subject to scurvy. But if you have access to vitamin C–containing foods, they are mostly all tangy and delicious. Vitamin C also plays in the big league of disease, as it is gobbled up by immune cells as part of their regiment for fighting infections. It's a natural antihistamine that dampens inflammation. This little job that vitamin C does on inflammation has positive implications in a variety of diseases, including cancer, atherosclerosis, acne, depression, asthma, celiac disease, IBS, and countless others. And vitamin C is also an antioxidant.

What Will Drinking Kombucha Do for Me?

Knowledge is not solely empirical; knowledge is also rational. In addition to the information we can gather from modern science, there is also a wealth of cultural wisdom we can draw from. From generation to generation over at least the last millennia and possibly the last two, kombuchasseurs have shared their bubbly beverage along with and because of their own experiences. The primary reason people shared their 'buch was to spread the health benefits they themselves had gained.

What follows is another list, a list of equal importance to the one above that was derived from empirical studies. This is the list of the health benefits that your fellow humans have experienced from drinking kombucha. This list is compiled primarily from people we know. We have added a few others that are cited in multiple publications.

The most illuminating views of kombucha's true health benefits arise from looking at the known effects of the compounds found inside your 'buch, in the list above, and comparing them with the list of ailments people have been relieved of from drinking it, in the list below. Spark some connections, drink some 'buch, and decide for yourself.

Drinking Kombucha Can Help Relieve

Acid reflux
Acne
Anxiety
Arthritis
Atherosclerosis
Colitis
Depression
Diabetes
Eczema
Excess weight
Fatigue
Fibromyalgia
Hangover
Headaches
Hypertension
Hypoglycemia
Indigestion
Premenstrual syndrome
Radiation poisoning
Rheumatism
Sluggish metabolism
Thinning hair
Tonsillitis

How can one beverage do all that? Until science has advanced far enough to be able to determine the complex pathways underlying many of these diseases, we may not know exactly. But that doesn't make the beverage any less amazing.

Kombucha can be viewed as a beverage that provides your body with the nutrients it needs to perform at the height of its ability at any age.

How Much 'Buch Is Good for Me? Who Should Drink 'Buch?

Moderation, folks. It's hard to hear when all you want to do is down a gallon of your new favorite health drink in an attempt to reap greater rewards. But really, everything should be taken in moderation. Too much of a good thing can become a bad thing. Even drinking too much water, the potion of life, can be deadly (although it takes *a lot* of water for this to occur). From our personal experience and what we've gathered from the experiences of others throughout the ages, we recommend that you start with four ounces of kombucha per day. After about a week at this level, you should know how the 'buch affects your particular body. If you want to, double it. Once you find yourself at sixteen ounces of kombucha per day, you should go ahead and make that your limit. With this volume you'll get all of the energy, detoxification, and nutritional benefits without the excess.

We also recommend taking some time off every few weeks if you are a daily 'buch drinker. Four weeks on, one week off is a good rotation, but pay attention to what works best for you. We teach this because, like we discussed earlier, too much of a good thing can be bad. Even good substances can start to build up in your body, and you want to give your body the chance to process those things too.

> Dependency is a yucky state. Be free creatures and don't let yourself be dependent on anything except the things we are absolutely dependent on: water, air, food, shelter, friends, and love!

Most people with a moderate to high level of health will experience nothing but amplified health when they start drinking 'buch! But pay attention to what your body is telling you. If you notice anything uncomfortable or off about your body when you try something new like kombucha, perhaps you should rethink doing it. Some effects that signal kombucha might not be for you are increased headaches, yeast blooms, or any other feeling of unwellness. When you first start drinking kombucha, you may notice that you are urinating more frequently or that food seems to move through your digestive system faster. These are normal effects that should go away once the initial detoxification has occurred, over the course of about two weeks.

If you have an illness you are struggling with, be mindful of the strong effects kombucha can have. Sometimes powerful nutrition can upset a system that is already struggling with imbalance. If you have any questions at all, please talk with your health advisors about kombucha and fermented foods in general.

Specifically, we're often asked about the interaction of kombucha with yeast infections. Yeast infections occur in bodies that have a compromised immune system for one reason or another. Another complicating factor is that there are several strains of yeast that can bloom in these situations that can be the cause of the infection. We have heard very positive reports of treating yeast infections with kombucha, and also some negative ones. We recommend that if you are fighting a war with yeast, try drinking a small amount of kombucha for a few days in a row and pay close attention to its effect. If you notice a beneficial change, keep going, and perhaps you may decide to drink a little more and pay attention to the effect of the increase. If you notice a negative effect, remove kombucha from your diet until your system has found balance another way.

Another common question is whether or not diabetics should drink kombucha. Regulation of blood sugar in diabetics is the subject of numerous studies involving kombucha. The studies are showing that kombucha does, in fact, have a positive effect on blood sugar regulation. But it is more complicated than that. For those sensitive to sugar, it is important to note that different kombucha ferments can fluctuate in the amount and kinds of sugars they contain. Individual diabetics have differing needs, with more or less flexibility in their sugar intake. Please talk with your health care provider to help you decide whether or not kombucha can be successfully added to your diet.

The History of Kombucha: Where on Earth (or Not Earth) Did It Come From?

Kombucha may seem new to many Westerners, but humankind has known about this wonder drink for millennia. So when you drink it or brew it today, you're connecting with an ancient tradition. We teach people to make kombucha using the original technology that has been used throughout the ages: fermenting tea and sugar using the kombucha culture.

The origin of kombucha is a mysterious one, with little concrete evidence to support the various stories that are told about it. Tales of kombucha's origin and history resemble myths that have been passed down through generations and have evolved through the ages, like a sweeping game of telephone, to meet the various cultural demands being placed on it at any given time and place.

For your reading pleasure, we have compiled some of the more exciting tales surrounding kombucha's origins and whereabouts over the ages. Take what you will from the past, as stories are always anchored in truth and history is subjective anyway. At the end of the day, whatever brings you closer to your ancient ferment is meeting the purpose.

What seems to be readily agreed upon is that kombucha originated somewhere in the Far East around the time that the leaves of the *Camellia sinensis* plant were first steeped into tea. The exact location, who discovered it, what was initially thought of the brew, and the rest of the details will perhaps always be elusive . . . until the fabric of time can be deliberately folded and someone can travel back to see for themselves!

QIN DYNASTY, THIRD CENTURY BCE

Emperor Qin Shi Huangdi was said to be the first emperor of a "unified China." His rule from 247 to 221 BCE, in the Qin Dynasty, was filled with achievements like the joining of the many sections that now make up the Great Wall to protect his empire from northerly invaders. As Emperor Qin Shi

Huangdi grew older, he began to fear death and became obsessed with achieving everlasting life. To this end, an alchemist gave him a magical elixir that he called "the drink of immortality" and promised to make him live forever. A few years into his kombucha experiment, the emperor died after trying out another path to immortality: ingesting pills containing mercury. Should have stuck to the 'buch, Huangdi!

DR. KOMBU, FIFTH CENTURY CE

The Dr. Kombu story is one of our personal favorites and a good story to get to the heart of kombucha. In 414 CE, a Korean doctor named Kombu-ha-chimu-kamu-ki-mu, known simply as Dr. Kombu, was summoned by the Japanese emperor Inyoko. The emperor was dying from a mysterious illness and was attempting to find a cure. Dr. Kombu was a revered physician who cared for his patients using a magical potion that he picked up in his travels throughout the farthest reaches of the Far East. Prescribing this potion, Dr. Kombu saved Emperor Inyoko's life. The Japanese were enthralled and celebrated the power of "Dr. Kombu and his magical cha." Kombucha, anyone?

JAPANESE SAMURAI, TENTH CENTURY CE

It has been said that the first use of kombucha was by early samurai warriors who would carry 'buch in their hip flasks to drink right before charging into battle. In this story, 'buch gave them the strength, cunning, and energy that the samurai were known for. It was largely responsible for the constant and definitive defeat of many foes. Try thinking about that next time you sip your 'buch in the morning.

TIBETAN MONK, LONG AGO

Another one of our favorites, the story of the Tibetan monk, illustrates kombucha's mystery and beauty. Long ago in the mountains of the Himalayas, a Tibetan monk was brewing a large batch of tea for his fellow monks. Before he could cover and put away the brewed tea, the monk fell into a deep nighttime meditation. During the night, gods who were about on the mountaintops blew dust down the slopes and into the window of the monk's home. The dust fell into the open

jar of tea and started to change it. When the monk completed his meditation, he discovered a gelatinous blob sitting on top of his tea. He passed this tea around to his fellow monks and they all lived strong and healthy lives.

ALIEN INVASION, 200,000 YEARS AGO

Some kombuchasseurs who are particularly open-minded explain the introduction of kombucha to this planet by an alien race. Delivered to prehistoric civilizations, this creature from another planet was ordered to be both eaten and drunk. Many might shrug off this opinion until they brew their first batch and see the alien-like kombucha SCOBY floating on the surface.

KOMBUCHA'S FAMILY TREE

Kombucha also has some close relatives in tibicos, kefir, some vinegars, and some ginger beers, which are all ferments that produce SCOBYs with each new brew.

Some people believe that kombucha evolved from one of these other SCOBY-producing ferments, or perhaps those others evolved from 'buch.

Whichever story rings true for you, it's clear that this effective elixir has been passed from aficionado to aficionado at a trendsetting pace, circling the globe countless times. Kombucha is currently experiencing a bloom in popularity throughout Western cultures as people embrace it as a healthful alternative to energy drinks and as a dairy-free probiotic. For those of you who remember the seventies, yogurt was just as uncommon and strange then as kombucha is now. Today it would be bizarre to find someone who doesn't know what yogurt is. We think kombucha is on that same track, and you are leading it there by joining the 'buch brigade.

However 'buch arrived on the scene, man, are we glad it made it to our party!

To further illustrate kombucha's world VIP status, here is a sampling of multicultural monikers it has left in its wake:

Cajnyj grib—Russian; translates roughly to "tea mushroom."
Hong cha jun—Chinese; translates roughly to "red tea fungus/mushroom."
Kocha kinoko—Japanese; translates roughly to "red tea mushroom."
Heldenpilz—German; translates roughly to "heroic mold."
Champignon de longue vie—French; translates roughly to "fungus of long life."
Miracle fungus—English; translates precisely to "miracle fungus."
Volga spring—English; comparing kombucha to the longest and largest river in Europe. Fittingly, the Volga River originates in Russia.

THE KOMBUCHMAN'S GRIMOIRE: ESSENTIAL HARDWARE AND TOOLS

Like warriors calmly preparing for battle or explorers heading out into the great unknown, kombrewers have essential tools and goods they will want with them on their brewing journey. Most of these items are already stocked in the standard kitchen.

A more advanced setup is easy to achieve if you take the time to get exactly what you need. The following checklist is your legend on the map to making your own healthy, tasty kombucha.

Brewing Hardware

Fermentation vessel (ferm vessel). Like a canvas to a painter, so is a fermentation vessel to the kombrewer. It is not just the holding tank; it is the vessel in which your 'buch will take shape, dance a microcosmic dance, and create life-affirming molecular compounds that will nourish your body. There are multiple vessel options:

Glass—A stock standard and an easy vessel to get your hands on. Glass provides both a window into the fermentation process and a safe no-leach environment for the 'buch to ferment in.

Pro: Glass is the perfect option for a beginner's brew and is very easy to control.

Con: Glass jars that hold more than three gallons are hard to come by and can be quite dangerous to work with and clean due to their extreme breakability.

Low to moderate cost.

Ceramic—Beautiful on the outside, clean and safe on the inside. A good ceramic crock can make the average home brew setup look like a professional fermentory. A wide variety of sizes are available. One-, two-, three-, five-, and ten-gallon crocks can be found at many local hardware stores, and we sell them on our website. When looking for a ceramic vessel, MAKE ABSOLUTELY SURE THE GLAZE IS LEAD-FREE. Lead-based glazes have the ability to leach during fermentation and can make your brew (or anything else that comes in contact with it) dangerous to drink. Most modern ceramics use lead-free glazes, but if you have any question at all, either test it or don't use it. Easy-to-use tests are also available at most hardware stores.

Pro: Ceramics have been around for centuries. They have fueled human creativity and furthered our progress as a civilization. Ceramics are beautiful, and your ancient kombucha ferment will be quite at home here. Ceramics are made of natural clay and are nontoxic if finished with a lead-free glaze.

Con: Ceramics are heavy. The bigger the ceramics, the more unwieldy they become. Oh, and did I mention to be sure to use nonleaded ceramic?

Moderate cost.

There are several awesome all-American companies that make ceramic vessels to order. Burley Clay and Ohio Stoneware are two of the American manufacturers whose crocks have been used in our brewery.

Plastic—Behold the mighty strength of the "light giant." Plastic has come a long way since Alexander Parkes introduced Parkesine at the 1862 Great London Exposition. Researchers have made astonishing advances in hydrocarbon technology, yielding plastics that can withstand all sorts of things. The plastics of today are able to tolerate the electrolytes, acids, and alcohols that are present in kombucha fermentation. But NOT ALL PLASTICS ARE FOOD-GRADE. There is a common misconception that all HDPE plastics (marked with the #2 recycling character) are food-grade containers. This, friends, is simply not so. Use only plastic containers that are *food safe*. If in doubt, ask the manufacturer.

Pro: Plastics are cheap, easy to clean, and virtually indestructible. You can recycle them! You probably already own stacks of food-safe plastic vessels waiting patiently in your kitchen for a new purpose.

Con: Our knowledge of plastics is evolving. Regardless of the density and type, there is still a chance that something yet unknown is leaching into our food from plastics currently marked food safe. If you can use something else, why not? It's better to be safe than sorry.

Low cost.

Stainless steel—From home brewer to pro brewer. If you are looking into stainless steel fermentation vessels, your love for the 'buch has clearly gone to the next level. The standard in beer breweries and wineries around the world, stainless steel is a beautiful and efficient vessel for your fermentation. Like plastics, many types of stainless steel are able to withstand low pH levels, electrolytes, and alcohols without leaching. Most beer and wine fermenting vessels are made of stainless steel types 304 and 316. Both of these have good strength and corrosion resistance. Beware of other 300-series metals, as their corrosion resistance is not up to snuff. Don't use any 400-series stainless steels—this is the type used in magnetic tools and is totally unsuitable for fermentation.

Pro: Sleek, easy to use, and easy to clean. Makes you look like a show-off, but at this level, you deserve a little fame.

Con: Expensive. Stainless steel vessels are the gateway to a more elaborate brewing setup. Once you start down this road, you may find yourself unable to turn back from a life of full-blown kombucha brewing.

Moderate to high cost.

Fermentation vessel cloth cover. Your 'buch needs to breathe. Oxygen and other atmospheric gases are essential fuel for the SCOBY during primary fermentation. Along with allowing air to enter your brew, your cloth cover allows the gases produced during fermentation to escape, so your brew doesn't suffocate under its own exhalation. If we lived in a sterile, bug-free world, no cover would be required. But since we don't, it is. When choosing a cover, find something breathable but with a fine weave. Old cotton T-shirts and kitchen towels are cheap (or free) and have an appropriately tight weave. Cheesecloth is an example of a fabric that is too porous and will potentially admit intruders like fruit flies, which can sneak in and occupy your brew. To allow for breathing and provide adequate protection, your cloth cover should be a few inches larger than the opening of your fermentation vessel. Secure the cover to your vessel's rim with a rubber band. Large rubber bands are available at office supply stores and online. Alternatively, just tie it on with a string. Seeing as we're all friends here, I'm going to let you in on a little secret: One year our ferm vessel was John McEnroe for Halloween, red headband and all!

Acetobacter strains present in 'buch brews feed off of oxygen in the air to consume alcohol, produce acetic acid, and participate in spinning the web of cellulose that holds a SCOBY together. Like a good yogini, kombucha unites the breath with the action!

Temperature gauge. Temperature is the wizard behind the curtain for most ferments. This is especially true with 'buch. It will dictate how long your brew takes to ferment and will play a big role in its final flavor profile. If the fermenting temperature is too cold, your brew may not progress properly and the potential for contamination by hearty molds will increase. Too hot, however, and your brew will become a sour vinegar bomb ingestible only by diluting it, cooking with it, or braving it. If the temperature varies widely over the course of the fermentation process, you will wind up with a mottled SCOBY that may not be as robust when used to inoculate new brews. In order to maintain a steady ideal temperature, you need to be able to monitor the temperature throughout the fermentation cycle. **Sticky thermometers** that adhere to the outside of your glass, stainless steel, or plastic fermentation vessel are the best option. Alternatively, and for ceramic vessels, use an ambient thermometer positioned a few inches away from your brew. Just make sure your thermometer is capable of accurate readings in the ideal 'buch brewing temperature range of 72 to 82°F (22 to 28°C) and a few degrees below and above so you'll know just how far you've strayed. If you don't have a sticky thermometer on the side of your ferm vessel, it is advisable to have an **instant-read thermometer** to make sure the temperature of your nute is within viable range before you add the SCOBY.

Stirrer or mixing spoon. Your spoon can be made from the same materials that are safe for use in your fermentation vessel, plus a few others. Since it seems a bit of overkill to use a giant glass swizzle stick anywhere outside a laboratory, may we suggest a wooden spoon for making kombucha. Wood is strong and long-lasting and takes on more character with each use. Wooden spoons are ubiquitous household paraphernalia that will be gentle on your glass or ceramic fermentation vessels for all you heavy-handed mixers out there. Beware the stainless steel mixing spoon in a glass or ceramic vessel! This can bring on a world of pain by creating a sticky, sharp mess.

Environment. In order for your microbes to operate at optimum capacity, they need to be placed in the right environment. For kombucha, an area that is room temperature (72 to 80°F) with good airflow is perfect. Many new homes have the luxury of central heating and cooling. If you are not in this group, don't fret. There are many ways you can keep your 'buch warm and toasty. Read on.

Heating element (optional). Brooklyn landlords are notorious for shutting the heat off during the coldest months of the year. If your place of brewing is anything like our kitchen, you will need some sort of heating element to keep your fermenting 'buch in the ideal range. Depending on your setup, there are a few easy, affordable options. The almost no-cost answer is using your oven as a low-temperature incubator. When just the light from the viewing lightbulb is on in a **closed oven**, it usually makes the perfect DIY incubator for fermenting 'buch. Test it out first by leaving a thermometer in your closed oven with the light switched to "on" for a couple of hours. Our oven keeps our ferm vessel at a steady 80°F during even the coldest of days with this setup. Just make sure to take out your jar before you fire up the oven! In fact, put a big fat sign over the oven dials so Auntie Jojo doesn't come into your kitchen and turn sugarplum dreams into hot dead kombucha.

If your oven is already host to some other culinary perfection, another inexpensive option is a **heat mat**. Low-wattage heat mats like the ones used in seed sprouting and reptile habitats can also keep your brew at the right temperature. A 5-watt heat mat is perfect for the job. Monitor your brew the first 12 hours to make sure it stays in a good temperature range. If the temperature starts creeping too high, you may want to move your vessel so that only a portion of it is resting on the mat, or invest in an inexpensive thermostat that can be used with the heat mat.

Perfectly sized and gauged thermostats and heat mats can be picked up from well-stocked pet stores in their reptile supply area.

Kombucha lets off a little bit of heat during fermentation. If your brew is just a couple of degrees cooler than you'd like, trap in that heat by wrapping a towel around it. You can do this in the oven setup, on a heat mat, or just by itself on the counter.

Heat source for brewing the nute solution. Water has to be heated to steep your tea. A stovetop or electric teakettle will get the job done quickly, but the sun can also do the job. Whichever method you choose, it's best not to bring the water to a rolling boil. In fact, it is best not to let your water stay hot for very long. The longer your water stays in the high temperature ranges used for brewing teas, the more the gases will escape—gases that could be better used by your microbial buddies in the SCOBY. If you are using a stove or electric kettle for heating your water, you will need a **heat-resistant pot** the size of your final brew volume to mix and/or heat liquids in.

Sun tea makes a very soft, clean kombucha. Highly recommended for floral, delicate teas like Dragon Well green or Silver Needle white. An elegant brew starts with an elegant process.

Measuring devices. You will need a measuring cup to measure your liquid 'buch starter and your sugar. For a 1-gallon brew, use a measuring cup with a capacity of at least 1 cup. A scale can be helpful if you are using loose tea instead of the pre-bagged varieties. Tea comes in different forms—whole leaf, balls, strips, powders, and so on. Measuring them by volume would give you widely varying amounts, whereas weight will give you accurate measurements every time.

pH indicator strips or pH meter. Although your taste buds are an excellent indicator of the doneness of your 'buch, you might want to go the extra step and get pH strips or a pH meter. On the budget-friendly side, pH strips will work just fine by changing color when they make contact with your 'buch. The color of the treated strip is then matched to a color on a pH chart that is usually included in the packaging. An electronic pH meter will give you precise pH levels but will cost you far more and is overkill for the home kombrewer. Once you get the hang of it, your taste buds will tell you what's up.

Airtight container. After your first brew is complete, you will have SCOBYs that will need protection while you bottle your finished brew and prepare your

next batch. Put your SCOBY along with your liquid 'buch starter in a container that tightly seals while you bottle your home brew and prepare for your next batch. If you are going to wait more than three days before setting up your next kombucha brew, store your SCOBYs and liquid 'buch starter in this container in the fridge. Glass is always recommended above plastic! But food-grade plastic works too.

Yeast filter (optional). At this point you need to decide whether you are going to pour your 'buch directly into bottles or run it through some sort of filtration. If you are one of those people trying to avoid yeast blobs or are trying to lower your yeast content to control alcohol levels (read more about alcohol in 'Buch Kamp II), you might consider filtration before flavoring or bottling.

In the basic setup, this would mean pouring the 'buch through a strainer, coffee filter, or cheesecloth. This will not remove any of the smaller yeast or bacteria cells, but it will pick up all the big conglomerates.

If you have taken your brew setup to the next level, either out of preference or to really control alcohol levels, then you probably have invested in some sort of filter cartridge housing and a pump to push it through it. This dense high-pressure filter removes quite a bit more of the yeast cells without affecting the bacterial content (unless you get really intense with your setup). This method is very effective, but it is expensive and complicated. Your local home brew shop can guide you through the various products that are available. Remember that FILTRATION IS NOT NECESSARY. Straight from ferm jar to bottle is the most common practice.

Yeasts are so tiny, they are measured in microns. To give you an idea of just how small they are, 1,000 microns = 1 millimeter, and different yeast varieties usually fall somewhere between 0.5 and 50 microns. You cannot entirely remove particles of that size from your brew with a coffee filter or cheesecloth. But then, you don't need to. More yeast equals more nutrients for you!

Auto siphon or funnel. When your 'buch is ready to bottle, you will need to use something to transfer it with minimum spillage and maximum ease. The economic, but often messy, option is to use a plastic funnel that is probably already in your kitchen and pour your 'buch slow and steady. The better option is to use an auto siphon. An auto siphon uses siphoning action to seamlessly move your 'buch from one container to the next, keeping it clean and your counters dry. It is essentially a simple hand pump and well worth the minimal investment.

Bottles. Kombrewers need to know what bottle will work best. The answer depends on what you are planning to do with your 'buch. Because flavoring also happens while you bottle, you will need to know how different flavorings affect your brew during secondary fermentation. For example, beet juice, chia seeds, and ginger have a lot of sugar in them, and until you get the timing right, you may wind up with an exploded bottle on your rack. It's best to try out new flavors in plastic soda bottles so you can monitor how much pressure is building in the bottle. Plastic bottles will get fat before they explode; you'll be able to halt secondary fermentation by putting them in the fridge well before that happens.

Once you know your mixture, glass is always best, and some glass options are better than others.

We have all gone through hundreds if not thousands of bottles by the time we are old enough to brew our own 'buch. Plastic water bottles, amber beer bottles, 16-ounce clear bottles of KBBK Kombucha, all of these can be washed with soap and water, rinsed, then reused for bottling your home brew. Many of my first bottles of home brew were repurposed 'buch bottles that I had picked up from the store. Just keep in mind that you need a good top that seals well to prevent the gases from escaping. Also make sure your cap will hold up under a little pressure. Small-mouthed soda, juice, beer, or water bottles are all fair game, just keep them in the single serving range so your 'buch doesn't go flat before you finish drinking it!

A Few Good Options for Glass Bottles

Swing-top bottle—A sturdy glass bottle with a little wire apparatus that keeps the top held down tight. A bonus of this bottle is that you can't lose the top! A good example of this bottle type is the Grolsch beer bottle. We've also seen fancy bubbly lemonade sold in this type of bottle.

Crown-cap bottle—This bottle is ubiquitous in the beer department. You will need new caps every time you bottle, and you will also need a crown capper. The whole setup is inexpensive. This is a good option for a regular home kombrewer.

Threaded-cap bottle—This is the most basic bottle type. Just make sure the caps you get for these bottles are able to withstand a good amount of pressure, as they often come with flimsy caps that are not built for this purpose. You may have to buy the caps separately, and remember: When in doubt, ask the manufacturer! We have a good variety of bottles with good caps for the purpose on our website.

Wine bottle with cork—If you drink wine, save the bottle and the cork. Cork has the ability to pop out if too much pressure builds up, but if you know how your brew is going to act in secondary fermentation, this setup will make a great presentation at your next 'buch party!

Bottle brush (optional)—For cleaning your bottles between uses.

Kombucha Ingredients

Now that we have gone over the hardware you will need, let's talk ingredients. Your choice of ingredients is fundamental to the flavor and health of your final product, so a little consideration is in order here. Your ingredients will first become the nutrient solution, or nute, that feeds your 'buch and will provide your SCOBY with the building blocks it needs to make the enzymes, acids, probiotics, and vitamins that your body is going to love it for. I like to remind people that your brew ingredients are primarily food for the culture, not you!

Stick with what works best for the 'buch. What works best for the 'buch ultimately means increased vitality for you!

Water. The silent, almost flavorless ingredient that plays a very large role in the health and taste of your kombucha brew. In most municipal water systems, chemicals like chlorine and fluoride are added to sterilize and, perhaps, make the water more nutritious. Whether or not chlorine and fluoride are good for you is up for debate, but we do know that those chemicals do not promote a strong, happy SCOBY. Although unfiltered tap water can be used without immediate repercussions, over time SCOBYs may suffer in this environment. Ideally, you will want to use filtered or bottled water. A standard carbon filter is a good start. May we also suggest looking into a ceramic candle filter? There are inexpensive countertop models that remove a whole lot of gunk from your water and make it tastier to drink plain, something you should be doing a lot of anyway.

SCOBY. Your guest of honor! Without this little happiness factory, the brew party would be pretty boring. There are many places to get a new SCOBY. Your friend next door may have an extra patty to give you, or you can buy a brewery-fresh SCOBY from us on our website. However you get your SCOBY, choose your source wisely. The microbial participants, genetic makeup, and overall health of your SCOBY will dictate the product it makes, both in flavor and in healthfulness. Make sure your SCOBY is an active culture that is fresh and coming from a clean, well-maintained place. Until you use it, keep your SCOBY in a sealed container bathing in some already brewed kombucha. If you are not planning on brewing with it immediately, put it in the fridge to slow down the metabolic rate—it's like a nice long sleep for your new friend. Only one SCOBY per gallon is needed for a successful home brew.

SCOBYs behave in a way that is quite similar to a multicellular organism. Having the good sense to take advantage of one another's strengths in a very competitive evolutionary playing field, the SCOBY community is organized in distinct yeast and bacterial layers that together have functions that exceed the sum of the parts. Drinking in the essence of pure synergy? Rad.

Liquid 'buch starter. Along with your SCOBY, your liquid 'buch starter is critical to a successful brew. Not only does it invite more microbial revelers to kick off your home brew party, it also immediately takes the pH down to a safe place for your SCOBY to live in. THIS IS A CRITICAL INGREDIENT TO A SUCCESSFUL HOME BREW. Never skimp on the amount of starter. Ideally, you will use already fermented 'buch as your liquid 'buch starter; add at least 1 cup per gallon of tea. If you do not have any kombucha around, 3 tablespoons of distilled white vinegar per gallon of nute will also do the trick.

Organic ingredients are higher in nutrients and lower in deleterious chemicals. They are also the most robust and best-tasting products on the market. Organic certification means that the farm has been scrutinized for its growing practices, which ultimately reward us with a lower environmental impact than conventional farming. That said, many tea varieties from East Asian regions are, in fact, organic even if they are not labeled as such. If in doubt, ask your dealer.

Teas. This is a complex and expansive subject. I could write an entire book on the art of choosing tea and another one on the art of brewing tea. Many people are scared away from getting too deep into kombucha brewing because they think it will require an extensive knowledge of tea to be a successful brewer. This is not the case. A general idea of the types of teas that are available and a good idea of what teas are best for the culture will give you perfectly complex 'buch. Heck, you can even brew with Lipton if it comes down to it and still end up with kombucha. But the best kombucha and the most delicate aspects on the palate are articulated by the type or types of tea you brew with. Also, the amount of caffeine in your finished kombucha will be determined by the amount you start with, as the caffeine content stays steady throughout the fermentation process. For more on tea varieties used in delicious kombucha brews, see 'Buch Kamp III.

Black tea—This would be the tea variety of choice if it were left up to the SCOBY. Black teas give your SCOBY a strong, nutrient-dense frame to work within. Black teas are high in nitrogen and tea tannins. Both supply the microbes with nutrients for fermentation and provide the fuel for strong cellulose production. Black teas take a little longer to fully steep and will give you deep woodsy flavor tones in your final brew in addition to the highest caffeine count. They also produce some of the strongest, thickest, most consistent SCOBYs you will ever see.

Ideal steeping temp: 200 to 210°F (just shy of a rolling boil)

Our favorites: ***English Breakfast, Yunnan Black, Darjeeling,*** **and** ***Golden Monkey.***

We typically blend our own tea by using a number of our favorite varieties to create a super brew for our family's SCOBYs to devour. KBBK's signature Straight-Up blend that we use in 90 percent of our kombucha production is a blend of specific green, black, and white tea varietals. Each variety plays a different role in the SCOBY's nute and contributes to the final flavor profile that we bottle and keg. Our SCOBYs love this blend, and it provides an awesome base in secondary fermentation.

Green tea—This delicate tea makes both a wonderful, mineral-rich final brew and a great addition when combined with others in a brew blend. Green tea kombucha ferments tend to finish a little quicker than black tea ferments and they produce a sweet, acidic back note on the palate. They are also lower in caffeine than black tea varieties. The cultures grown solely on green teas are slightly weaker/thinner than those grown on black teas and are more transparent.

Ideal steeping temp: 170 to 180°F

The difference between black tea and green tea, both from the plant species *Camellia sinensis*, is that green tea is only minimally oxidized, goes through very little fermentation, and is dried quickly in production. This seals in some extra health benefits for the drinker, most

(continued)

notably in the way of highly concentrated, unoxidized antioxidants called catechins. Catechins bind and dispose of free radicals in the body that cause DNA damage that can result in various cancers, blood clots, and atherosclerosis. Green teas also contain less caffeine than black tea.

Our favorites: *Dragon Well*, *Jasmine*, and *Clouds and Mist*.

White tea—The youngest buds picked from the plant, white teas are simple, delicate, and very low in caffeine. It's often the most expensive tea you can get and has a very clean and almost sweet flavor profile. In a 'buch brew they provide some natural fruity sweetness that can make even the haters change their mind. Try brewing your white teas using the power of the sun.

Ideal steeping temp: 160 to 180°F

White teas are picked earlier in the tea plant's life cycle than green and black varieties. They are allowed to wither in natural sunlight before very minimal processing, resulting in low caffeine and high catechin counts.

Our favorites: *Shu Mee, White Peony,* and *Silver Needle.*

Oolong tea—A traditional Chinese tea, oolongs undergo a unique and time-consuming process that includes drying under the sun, heavy oxidation, and, in many cases, a curling and twisting technique that gives the finished tea leaves a ball-like characteristic. Oolongs can be very expensive but yield some of the most amazing kombucha brews ever to grace the lucky palate that gets to taste them. Oolongs give the culture a unique blend of nutrients, yielding a robust SCOBY and a balanced acid profile. Flavor characteristics vary among the different oolongs, from some that impart woody flavors to others that give bright honey-like flavors. The names of these teas are closer to the title of a Bruce Lee film than a dainty tea bag. Bad-ass!

Ideal steeping temp: 190 to 205°F

In order to make an oolong supply last, use it in a blend with other teas. See note below in blending teas.

Our favorites: *Red Robe, Phoenix Mountain, Golden Jade,* and *Iron Goddess.*

Single-origin teas are an ecological marvel to brew with. A kombrewer can home in on specific flavors and attributes that they are drawn to, as well as experience flavors distinct to particular growing regions and seasons. A really sensitive kombrewer will also notice that their SCOBY will behave differently with teas from different regions. It's a fascinating study and one that's not any more costly than brewing with the commercial blends.

Sweetener. Your SCOBY's high-octane best friend. Most SCOBYs that fall into the hands of a kombrewer will do best when fed pure white cane sugar. The vast majority of commercial kombucha producers use pure cane sugar in their nute, as do the vast majority of kombucha home brewers. Growing awareness of the health hazards of refined sugar has led new brewers to ask if they can use alternative sweeteners to feed their SCOBY. The first part of the answer is that the sugar is for the SCOBY, not for you. Pure refined cane sugar contains nothing but sucrose, an instant fuel source for hungry microbes to use in making all of the different compounds we listed in Chapter 1. Refined sugar has no fiber, no fat, no protein, no sodium, no calcium, no potassium, no antibiotic compounds, no complex sugars—nothing to stand in the way of the microbes getting to their favorite fuel, sugar molecules. If you are looking to feed your culture the simplest food, pure and simple sugar is the way to go. And for you, fully fermented 'buch has as little as 1 percent of that sugar left when you drink it.

For the sugar-conscious kombuchasseurs out there, let the ferment go until all of the sweetness is gone. When your 'buch tastes all sour and no sweet, you'll know the sugar is pretty much nonexistent. Blend it with some unsweetened tea and you'll have an awesome-tasting low-sugar beverage!

The second part of the answer is yes, you can brew kombucha using alternative sweeteners, but you will have to either train your SCOBY to eat something new or find a SCOBY that someone has already trained to your sweetener of choice. By switching sweeteners, you are essentially changing the nature of the ferment. This must be done slowly so that you don't shock your culture into lifelessness. It will take some time for your SCOBY to "learn" to feed off a different nute. By training your SCOBY to an alternative sweetener, you are teaching it to work with different building blocks and overcome different hurdles. The microbial population of your SCOBY may change, as will the components those microbes produce, thus altering the kombucha ferment. It's like making wine from watermelon juice. The juice will ferment, but you are going to get nothing close to a fine Italian Barolo.

A SCOBY that feeds off of honey instead of cane sugar is called a Jun SCOBY. Train your cane sugar SCOBY to eat honey by adding 1 tablespoon per brew along with the sugar, then adding more honey and less sugar on each subsequent brew. Just keep in mind that honey contains antibacterial agents that may deteriorate the culture over time.

Some of the best 'buch results from a little experimentation. Depending on the alternative sweetener you choose to train your SCOBY to, over time your SCOBYs may become less and less robust. No worries. Every good scientist knows to hold on to some of her original colony in case something goes wrong. Always keep a SCOBY or two fermenting in a nute of your "house blend" so that you can re-culture if a test batch goes awry.

We know people who have successfully trained SCOBYs to molasses, demerara sugar, honey, maple syrup, and agave. Kombucha utilizes glucose, sucrose, and fructose. Therefore, kombucha cannot be made from stevia, xylitol, lactose, or any artificial sweetener.

Flavorings (optional). To learn more about flavoring ingredients and technique, read on! We talk about flavoring in 'Buch Kamp II and III. The options are endless; just remember that organics taste better and floral qualities are your kombucha's best friend.

’BUCH KAMP I: BREWING AND FERMENTATION

Wait. What’s ’Buch Kamp?

Attennn-tion!

Now that your kitchen is stocked with the tools you will need to hone your skills as a kombrewer, it is time to enter ’Buch Kamp! In ’Buch Kamps I, II, and III, you will go through basic training to prepare yourself to fearlessly create healthy and delicious kombucha brews that will add the good to and take away the bad from your body, and perhaps your life. What was once one of the great mysteries will be revealed and you will become master over microbe! Or, as we like to call it, a ’buch brigadier. Read mindfully and you will come out of this chapter a confidant, strong kombrewer.

The skills you are about to learn in ’Buch Kamp will not only help you master ’buch creation, but will also give you a new way to look at your food. It is the start of a magical journey into the art of fermentation. Once you understand some of the basic principles of making kombucha, you will begin to see food in a whole new way! It's time to join the kombucha revolution and take control of your food.

Brewing Instructions

'Buch Kamp I focuses on the art of brewing and primary fermentation. It is the beginning of the saga and the foundation upon which everything is built. Read this section thoroughly. If at any point you get lost, e-mail us at HomeBrew@KombuchaBrooklyn.com, or connect with us on any of our social media platforms. We want you to brew with confidence!

SETTING UP YOUR BREW

All fermented foods begin with a nutrient-dense substrate, or nute, that will undergo transformation by microorganisms. A nute is basically an awesome microbial habitat in which organisms can live and have their ravenous appetite satisfied. In the case of kombucha, the nute is water that has been steeped with black, green, and/or white tea and sweetened with a mega dose of sugar. That blend becomes powerful food for the kombucha microorganisms that you will purposefully introduce to the mix. I say purposefully because some ferments are transformed by microorganisms that are naturally present in the substrate, as is the case with sauerkraut or wild sourdough bread.

For kombucha, the microorganisms that you will purposefully introduce are in your SCOBY and liquid 'buch starter. Kombucha ferments contain up to thirty strains of yeast and bacteria. Your SCOBY will eat and ferment your sweetened tea nute, transforming it into the complex cocktail of acids and enzymes that we have all come to love. Starting with a solid nute is essential to a healthy, tasty finished product. Therefore, take the utmost care when making it.

The job of the kombrewer is not to make the kombucha. The kombrewer's job is to curate the environment in which your SCOBY will flourish and provide you with robust kombucha. Mix and measure everything with attention to detail. Create a nourishing environment for your SCOBY to thrive in and it will give you some seriously amazing 'buch.

Before you dig in, clean your workspace. Dish soap is perfectly sufficient unless you have been cutting your poodle's toenails or something on your kitchen counter. In this case, we would recommend using something a little stronger and perhaps antibacterial. This will get rid of all the non-kombucha microorganisms that might be lingering on the surface. After that is done, be sure to rinse off any soapy residue with water. You don't want your SCOBY to come into contact with soap or antibacterial sprays that could harm your hardworking culture.

After your surface is squeaky clean, give your hands the same cleansing treatment. Rinse, rinse, rinse!

Next, prepare your mise en place. *Mise en place* is a French phrase used in kitchens that refers to prepping all your stuff, measuring your portions, and having everything ready in an accessible location. All of the equipment and ingredients that you gathered in the previous chapter need to be clean, readily accessible, and organized on your work surface, so lay out your implements and ingredients in a way that makes sense. I like to arrange them in the order in which I will use them. Once this process is in motion, it's best if you are not running around outside your clean brew zone looking for that fermentation jar. Contamination is not a high risk, but it is always there. Why not prevent against it?

Your Brewer's Checklist, in Order of Appearance

Clean side towel (for mopping up spills and such)
Water
Water heating apparatus (pot and stove, electric kettle, pitcher
  for sun tea, etc.)
Instant-read thermometer (if not using a sticky thermometer)
Mixing pot
Tea
Dry measuring cup
Sugar
Spoon
Fermentation vessel
Liquid 'buch starter
SCOBY
Cotton vessel cover
Rubber band
Liquid measuring cup

The amount of water, tea, sugar, and liquid 'buch starter will vary depending on the size of the fermentation vessel you are using. Below we list the amount of each ingredient used for a one-gallon brew. For a half-gallon brew, cut it in half. For a two-gallon brew, double it. For a four-gallon brew, quadruple it. Get it?

This recipe is totally scalable . . . to a point. Don't increase to a fifty-gallon brew until you really know the ins and outs of kombucha brewing. There is a surface to volume ratio that must be maintained when you get into large-scale brewing.

> Sanitization is not necessary, but we HIGHLY recommend that you keep a bottle of overfermented 'buch vinegar close to your workspace. Rinse your brew vessel with this before you add your brew. A cup of 'buch vinegar sloshed around in there and then poured out will ward off any microbial intruders and prep the environment. Kombuchasize it! Another option is using a specialized fermentation sanitizer that can be purchased at your local home brew shop or online.

Kombucha

Makes 1 gallon

⅞ gallon (14 cups) cool filtered or distilled water
6 standard-size tea bags or 13 grams of loose tea, bagged
1 cup cane sugar
1 cup liquid 'buch starter or 2 tablespoons distilled white vinegar
SCOBY

1. The first step in brewing is to heat about one-quarter of the water you intend to brew with. For our 1-gallon example, heat ¼ gallon, or 4 cups, water. If

you know what temperature is ideal to steep your particular variety of tea at, keep an eye on the temperature with an instant-read thermometer. Some countertop teakettles have temperature settings you could utilize. If you do not have a tea that is suited to a particular steeping temperature, just bring your water to a boil, then shut off the heat source. Most teas do fine at temperatures just shy of boiling.

As temperatures rise, the gases that are dissolved in your water, like nitrogen and oxygen, will start to escape into the air. Bring your water to the correct temperature for tea steeping, then shut off the heat source ASAP. That way, more gases will still be there when the microbes join the party.

For sun tea, put three-quarters of your water, or ¾ gallon for our 1-gallon example, into your pitcher along with your tea bags. Cover the pitcher and set in a sunny place for up to 6 hours. Remove the tea bags. Mix the remaining water and the sugar together until the sugar completely dissolves. You may need to heat this water to get the sugar to dissolve completely. Combine your sugar water with your sun-brewed tea in your ferm vessel. Now jump in with the rest at step 6.

2. Once your water is at the correct temperature, add your tea bags or diffuser containing loose tea. Give the blend a good stir, making sure the water has saturated all of the tea. Use your spoon to press and release the tea bags

against the side of the pot to extract more of the tea goodness before allowing it to sit, covered.

Every 5 minutes, repeat the stirring and pressing process. This will ensure that all of the nutrients are steeped from the tea. Repeat this 3 times for a total steep time of 20 minutes.

You're looking at your tea and saying booyah! That strong brew will turn your hairless sphinx into Cousin It! The point of making such a small, strong batch of tea is to save you time. One gallon of hot tea takes hours to cool to a temperature that would be safe for your SCOBY to be introduced to. By steeping strong and then diluting it with cool water, you will immediately drop the temperature to the proper place and save yourself some time.

While your tea is steeping, put the remainder of your water in the fridge. This will ensure that when the time comes to combine the two, you won't be waiting around twirling your thumbs while your nute cools to a temperature that won't kill your SCOBY.

3. After 20 minutes of steep time, remove the tea bags, giving them a final squeeze. Every nutrient-filled drop counts.

4. Add your premeasured 1 cup of sugar to the hot, hefty tea. With your spoon, mix up the solution

until you don't see any more sugar crystals in the pot. Don't be afraid! Remember ALL OF THAT SUGAR IS FOR THE SCOBY! Not for you.

5. You now have a small amount of very strong, very sweet tea in a large pot. The next step is a critical one to get right. Many new brewers forget to do this in the right order, and then a week later we get an e-mail saying, "Nothing's happening!" Well, duh, cowboy. If the solution is still hot when you put your SCOBY and liquid 'buch starter in, it will kill it outright. So the next step, 'buch brigadiers, is to dilute your hot, strong tea with the remainder of the cool water you have. This will immediately bring it into a safe temperature range for your little buggers to *thrive.*

6. Once you've added the cool water to your nute, the solution should be cool enough to go ahead and put it in your final fermentation vessel. So do that. Pour your nute into your clean ferm vessel.

7. Take a step back and look at what you've done. In your ferm vessel is almost 1 gallon of sweet tea. That is now your finished nute, ready to nourish your SCOBY and be transformed into the beverage of the gods: kombucha. There is only one more thing you need to do before moving on to the next step:

   CHECK THE TEMPERATURE OF YOUR NUTE BEFORE GOING FORWARD.

   Did you hear me, brigadier? I said, look at that sticky thermometer or stick that CLEAN instant-read thermometer into your fermentation vessel and make sure the temperature does not read above 90°F. If it does, you will have to cool your haunches until the temperature drops to below 90°F. Cover your vessel if you have to let it sit.

8. Once you absolutely know that your nute is cool enough to sustain and not destroy your kombucha microorganisms, add the liquid 'buch starter. That starter is what's going to bring the nute to the proper pH range that will protect your brew from foreign microbial invasions and jump-start the activity of your SCOBY.

9. And now . . . the moment you have all been waiting for . . . at long last, get out your SCOBY! It's time to set it free in its new home. Gently place (plop? drop? splash?) your SCOBY into the ferm vessel, where it will float, sink, or hover in between for the next week or two.

10. With your SCOBY floating happily in its new home, secure your ferm vessel cover to the rim with your rubber band. Hollah! The 'buch is fermenting! Now it's time for the real work to begin . . . with you sitting comfortably on the porch . . . or snoozing in the hammock with a cocktail. Set the ferm vessel in a location where the right temperature will be maintained. Don't put your ferm vessel in direct sunlight, mainly because the temperature won't be easy to monitor and

maintain, but also because sunlight is also known to diminish nutrients. The ideal kombucha fermentation temperature is above 68°F and ideally between 72 and 82°F.

The torch passes from kombrewer to komfermenter. Let the fermentation begin.

THE FERMENTATION PROCESS

SCOBY formation: As your brew ferments, you will notice changes in the nute. Most notably will be the formation of the new "baby" SCOBY on the surface. This process begins in most brews between twenty-four and seventy-two hours. Small white patches will begin to form on the surface of the liquid, independent of the SCOBY you put in there. The first few days are an uneasy time for new brewers, and the new growth of SCOBY is often misconstrued as mold. We talk about mold in greater depth later in this chapter. Check in there if you are unsure of what you are looking at. And don't hesitate to e-mail us photos if you'd like a second opinion. As the culture matures, these spots of new growth will become thicker and wider, eventually joining together and becoming one whole patty.

In his book *Permaculture: A Designers' Manual,* Bill Mollison writes that transition zones—the places where field and water, row and hedgerow meet—have greater biodiversity than the middle. It's only fitting that the SCOBY would grow at the border between air and liquid proliferating in the transition zone.

Taste your brew: When the new baby SCOBY has spread across the entire surface of your brew and started to thicken, you should give your 'buch a taste. This will usually be in the three- to six-day range but can take longer depending on the strength of your culture, how long you have been brewing in that location, the type of tea and sugar used, and the temperature. Lots of changes have already occurred in your brew at this point, and the flavor will give you an idea of how much longer you will want your brew to ferment. Just make sure that if you dip something into your fermenting 'buch, it is clean. Kombucha goes from sweet tea to delicious 'buch in about two weeks and from delicious 'buch to 'buch vinegar in another week or so. As long as your brew is healthy and progressing normally, it's always safe to drink from the nute stage all the way through to vinegar.

Some ideas on how to get a sip:

- Stick a straw under the surface of the SCOBY and suck on it.
- Use a clean shot glass to gently push the SCOBY down and scoop a little from the surface.
- You can even pick up the whole jar and tip it back to your lips. What better way to say hello to your new SCOBY

friend! By the time your brew is covered with new SCOBY growth, it is acidic enough that your lips can't contaminate it anymore. You're in the zone.

Yeast strands and blobs: Along with the new SCOBY, you may start to see brown strands forming in the brew. They might hang from the SCOBY or float freely in the nute. These strands are yeast colonies that like to stick together in long chains. Don't fret; they aren't some weird unwanted party crashers. They are a part of the brew and offer you nothing but kombucha goodness. If you want to minimize these strands in future brews, filter them out of your liquid 'buch starter before you inoculate your brew (more on this in 'Buch Kamp II). This will limit their growth. Sometimes the yeast will also attach itself to the top of the SCOBY, form blobs, and even dry on the surface, giving the culture dark brown spots or areas that many new brewers easily confuse with mold. We talk more about mold toward the end of this chapter.

When to bottle: Most one-gallon brews kept around 78°F will have a nice balance of sweet and sour flavors at nine days. I like to bottle at about seven days in my kitchen when there is a little more sweetness than I would want to drink. This ensures that there will be enough sugar to produce effervescence in secondary fermentation after bottling (secondary fermentation is covered in 'Buch Kamp II). However, sometimes I choose to bottle it on the dry (less sweet) side and let the brew go up to fourteen days. I do this when I want to be able to add sugary flavorings during secondary fermentation. Temperature will be the primary controller of when your brew will be ready, and your palate will be the judge.

A pH indicator measures the activity of hydrogen ions in a solution. The more free-floating hydrogen ions there are, the lower the pH will be, indicating a higher acid profile. For the kombrewer with pH indicator strips, your 'buch will be ready on the sweet side at a pH of 3.1 and on the sour side at a pH of 2.7.

If you let it go longer and it becomes so sour it tastes like vinegar, no worries. It is 'buch vinegar now! You can use it in some of the tantalizing dishes in Chapter 6, or give yourself a beauty treatment using one of the recipes in Chapter 8. Or water it down and use it for a healthy, totally nontoxic house cleaner!

RECAP

Your brew has come a long way since we started. It has gone from hot water to sweet tea to detox tea in only a matter of days, while at the same time propagating itself for use in future batches. This complex process was facilitated by you, the kombrewer, and made possible by the millions of microorganisms in your SCOBY and the generations before you who passed the culture down through the ages. Behold the magic of fermentation.

Now that you have a batch of 'buch, it's time to do something with it. Many brewers go straight from ferm vessel to cup, an awesomely simple way to rock your 'buch. If you are one of those kombrewers, put your brew in a sealed container and keep it in the fridge. This will keep your 'buch from turning into vinegar. If it stays within the fermentation temperature range, it will continue to ferment.

If you are a little more adventurous and can wait a few more days before digging in, let's take your 'buch to the next level! Follow me and let me introduce you to secondary fermentation. We will explore secondary fermentation and flavoring in 'Buch Kamps II and III.

Komplowing into the Future: Your Next Batch

You've made and are about to bottle your first batch of 'buch. Well done, cadet! You've come a long way. Now that your first batch is in the books, it is time to start a second batch. Do you need to go out and source another SCOBY? Nope! Check it out: There is something new in your ferm vessel. A "baby" SCOBY was born in your previous fermentation round and new life has emerged.

Kombucha fermentation is amazing, isn't it? Like many other ferments, the new life that has taken shape in your first batch of kombucha can be used to start your next round, and so on. The special difference with kombucha as opposed to most other ferments is that this new life is a solid physical object that you can pick up and handle. Your new SCOBY that grew on the surface of your first batch is an awesome display of life and microbial power. Now go ahead, feel it. You know you want to. Don't be scared, it won't bite. Or will it . . . *mwah-hah-hah!* Just kidding. It's slimy and fun! Kids love them.

Your next one-gallon batch will need only one SCOBY and one cup of your previously fermented 'buch to use as liquid 'buch starter. Just follow the instructions all over again using your newest SCOBY and your saved liquid 'buch starter.

Don't get attached! Your brew jar can quickly become a SCOBY hotel if you're not careful. Unless your goal is selective breeding or banking, regularly give away, compost, eat, or up-cycle your new baby SCOBYs. This book is full of tasty alternatives. Read on!

Foreign Invaders: Unwanted Kompany

Of the ferments that one can make at home, kombucha is one of the easiest to keep safe and healthy. Much of the health of your brew is predetermined by your liquid 'buch starter. When added to the nute in the proper amount, liquid 'buch starter will immediately lower the pH. This gives your fledgling batch of 'buch an instant competitive advantage over unwanted microbial contaminants like mold or harmful bacteria.

When a problem does arise, however, it is good to know what you are looking at. The most common foreign invaders are listed below. If any of these happen,

toss your entire brew, SCOBY and all, and start over from scratch. These guys will weaken and eventually destroy your brew, not to mention that they may change the flavor and just flat-out make you not want to drink the stuff. It is important to recognize that it is difficult to make kombucha in a way that will actually make you sick. Read on and learn about kombucha's competitors.

Mold. The most common contamination is mold. Molds are resilient microscopic fungi that can grow on almost any type of nutrient-dense material. We've all let a loaf of bread go a little too long and had greenish, bluish, whitish, fuzzy, soft mold grow on it. It's generally not dangerous unless you have an allergy to mold. Some people just cut it off and carry on with the sandwich making.

Just like bread, when your 'buch gets infiltrated and overtaken by mold, you will know it. A patch of either green, blue, white, or black fuzzy stuff will be growing on top of your new SCOBY or on the surface of the liquid. Many first-time brewers will mistake a yeast swell or new SCOBY formation as mold. The way to overcome these freshman blues is to look at it like a taxonomist. Mold needs air, so it will always form on the surface of the SCOBY or liquid. It will never form underneath the surface of the liquid. It will either be dry and fuzzy on the surface or look like a separate distinct entity growing in splotches that don't really meld together.

When in doubt, look online! There are countless websites (including ours) with image libraries of mold-contaminated SCOBYs. Compare yours with the images and pay special attention to the points I just mentioned. At the end of the day, if you are still unsure, e-mail a picture to us! We are always happy to help.

If you confirm that you do have mold, don't futz around. Toss the entire batch and start again. If you have other brews next to a mold swarm, it can easily spread through airborne mold spores, so go ahead and toss those too.

Mold contamination is usually caused by culture death from adding the culture to liquid that is too hot, under acidification when you set up your brew, or fermentation temperatures that are below the optimal brewing range.

To Beat Mold

- Always verify that your nute is below 90°F before adding the SCOBY or liquid 'buch starter.
- Always keep your fermentation temp in the right range.
- Always add the correct amount of starter to your nute.
- Avoid letting any particulate from your tea into the fermentation vessel.

Fruit flies (*Drosophila melanogaster*). These little devils come out of nowhere. Sometimes it feels like they appear out of thin air. They are little warriors that somehow find their way into your brew with even the most careful precautions. Fruit flies like to make a nice home for themselves and their burgeoning families on your SCOBY. Once there, they will immediately get frisky and start multiplying, making little baby flies, aka larvae, that will creep and crawl all over the surface of your brew. Those little baby devils will eventually hatch and become flies themselves, trapped under your fermentation vessel cover. If left unattended, one fruit fly will create an empire of brothers and sisters claiming *your* home brew as their own. If you get an infestation of fruit flies, toss the entire batch and start again. You may be a 'buch brigadier, but dude, you are no match for an army of fruit flies.

Fruit fly traps are easy to make, and might just be the difference between tasty brew or no brew. An empty small-mouth bottle with a paper funnel stuck in it and a splash of kombucha in the bottom will attract flies into entrapment and eminent doom. In the summer months we catch thousands that way.

Vinegar eel nematodes (*Turbatrix aceti*). Vinegar eels are an interesting breed. Commonly found in vinegar manufacturing facilities, they feed off of the microbes that proliferate in both vinegar and kombucha. Vinegar eels can also be found in the ocean and in acidic lakes, so if you've ever been swimming in the rough, you've probably rubbed up against these guys at some point.

Although uncommon in the home brew setup, if you do get them, vinegar eels are difficult predators to detect. A brew of 'buch can house vinegar eels for months, and only the trained eye during careful review will notice them. They are passed down from batch to batch, and it takes a while for the effects to become apparent. The good news is that the FDA has determined they are harmless and nonparasitic. The bad news is that over time vinegar eels will slowly kill your culture. If your kombucha is not developing as quickly as it normally does, if the flavor profile changes unexpectedly, or if your SCOBY starts to deteriorate and you find it lacks structure, check for vinegar eels.

Vinegar eels can be seen with the naked eye, but only if you are really looking. Find them by taking your brew into a dark room and shining a flashlight into the ferm vessel. If it's clear glass, you can shine your light into the side. If you are in ceramic or stainless steel, shine your light into the top after pushing the SCOBY way down. If you've got eels, they will swim and wiggle their way toward the light. If vinegar eels are discovered, toss all contaminated brews and start fresh with new SCOBYs and liquid 'buch starter from noncontaminated sources. Don't mess around when you are cleaning up after a vinegar eel infestation. It is remarkable how resilient they are. Use boiling water to sterilize everything that has come in contact with contaminated brew before starting a fresh batch.

It is always best to restart with a fully formed SCOBY, but if you are desperate, a successful batch of kombucha can be started with just a bottle of kombucha and your nute. All the microorganisms and yummy goodness that is needed to start a new kombucha SCOBY is in the kombucha itself. If you start a new brew from just a bottle of kombucha alone, your brew will take an additional one to two weeks to fully ferment. The first SCOBY your new brew makes may not be the most robust SCOBY in the world. In time your culture will adapt and gain strength. A few batches down the line and you will have a nice culture going, SCOBY and all. Ferment in the upper region of the temperature, around 80°F, if using only liquid 'buch starter.

'BUCH KAMP II: BOTTLING AND SECONDARY FERMENTATION

Let the games begin! Just when you thought you were a kombrewmaster, along comes your neighbor with a growler of her home brew, and dagnabbit, it blows yours clear out of the water! When she opened that fine amber bottle, a seductive hiss of gas and a suggestive dollop of white foam came out. A splash across your tongue yields the perfect prick of effervescence. And that's when you ask the million-dollar question: *How do I get bubbles in my 'buch?*

I'm here for you, 'buch brigadiers. 'Buch Kamp II is all about what to do after your kombucha has fermented to take it from ordinary to extraordinary. It all starts with bottling your home brew. What type of bottle you use is very important to secondary fermentation, the process in which yeast creates gases in your 'buch. The SCOBY has done its job and made you kombucha. Now it's time for you, the kombrewer, to master one of the most difficult but rewarding beasts in the world of fermentation—yeast. Let's get started bottling your home brew and we'll talk more about this as we go along.

Mise en Place

Additional flavorings (optional). Some great flavoring suggestions are laid out for you in 'Buch Kamp III. If you plan on flavoring, make sure you read the rest of 'Buch Kamp II and 'Buch Kamp III before you bottle your brew so you can get the full picture. Of course, flavoring is totally optional; you can make a damn fine beverage just using your Straight-Up brew.

If you are flavoring your kombucha, bottling is when you will add your flavoring ingredients. Some flavorings can be mixed into the 'buch before transferring it to bottles. Examples of these would be juices, purees, and powders—flavorings that mix right into the solution and either dissolve or are held in a homogenous suspension. Flavorings that will be added as larger particles, like fruit (chunks, dehydrated, or freeze-dried), slices of roots, or whole herbs, should be added to each bottle individually.

Flavoring your 'buch bottle by bottle is a great method no matter what you are flavoring with. It takes a little more work, but the reward is the ability to create multiple distinct masterpieces with each brew. Individual flavorings also allow you to try different concentrations of the same ingredients. It's amazing how the subtle aspects of flavors can be brought out by adding just a little less or a little more of them.

Yeast filter (optional). At this point you will decide either to filter or not to filter your 'buch. It is really a matter of preference. It will have no bearing on the health of your 'buch or the level of effervescence you can achieve. Filtering off the yeast will clarify your kombucha, making it picture perfect. It may also decrease the speed and degree of secondary fermentation achieved. If you have decided to filter, you will want to do it before adding your flavorings. You will also need a second container to transfer your 'buch into during the filtration process.

Ferm vessel full of ready-to-bottle 'buch. You determined your 'buch is ready to bottle using the guidelines in 'Buch Kamp I. If you've decided to bottle it slightly on the sweet side, secondary fermentation will be a lively event. If you are bottling it on the sour side but still want some bubbly kick, perhaps you want to flavor it with something sugary, like raisins or beet juice, so the yeast will have lots to munch on and make that delectable CO_2.

If you are mixing in a flavoring ingredient before transferring to bottles, you will probably want to just go ahead and mix it all up in your ferm vessel. Hey, why make more dishes if you don't have to, right? *Before mixing in any flavorings, be absolutely sure to remove the SCOBYs and enough liquid 'buch starter to start your next batch.*

Airtight container. Store your SCOBY and liquid 'buch starter in here while you bottle and prepare for your next batch. If you are going to wait three or more days before brewing again, store them in the fridge. SCOBYs can cohabitate with other SCOBYs in the same container. Just make sure there is enough liquid to completely submerge them.

Bottles. Whatever bottle type you choose, be sure to clean them with soapy water and rinse them thoroughly before bottling your 'buch. A good bottle brush will come in handy.

If you want to monitor your secondary ferment, bottle all your flavored brews in glass except for one, which you'll bottle in plastic as a tester. Upon bottling, the plastic bottle will be squishy to the touch. As pressure builds, the bottle will become harder and harder. When the plastic has no give at all, you will know your other bottles are ready to put into the fridge without having to open them, preserving their yummy bubbly goodness.

Funnel or auto siphon. Whether you've mixed in soluble flavorings or are bottling your brew in the buff, these are the two ways we recommend transferring your 'buch into bottles. If you have an auto siphon, try out your technique using water in the sink before you go for the gold with your 'buch. You'll thank us for making you practice.

Cleanliness! Whether you use a funnel or an auto siphon or you are just hoping you aim true, it is essential that you keep the process clean and free of foreign contaminants. Before you begin: Clean everything! Just as in 'Buch Kamp I when you were preparing your batch for primary fermentation, it is essential that you keep your 'buch clean and free of foreign contaminants during the bottling process. Soap and water clean well enough. And throughout the process, *beware of the crafty fruit fly!* As soon as you start splashing that sweet golden liquid around, they will come. Trust us, they will come. Even if you live in Antarctica.

Bottling and Flavoring Instructions

1. **Optional filtration:** If you are going to filter, you will want to do it before you flavor and before you transfer your 'buch into bottles. Whether using a funnel or an auto siphon, you want to pass your home brew through your filter or strainer into a second container.

2. **Flavoring with soluble ingredients:** If you are using flavoring ingredients that can be mixed into your entire batch, go ahead and do so now. For specific recipes and amounts, look forward into 'Buch Kamp III, the section of 'Buch Kamp devoted to the higher arts of flavoring.

3. **Transfer to bottles:** After you have prepared your 'buch for bottling, use your auto siphon or funnel setup to do so. If you will be flavoring bottles individually, keep in mind how much room you will need to leave for those additions.

4. **Flavor individual bottles:** If you have decided to add flavorings to individual bottles, either because you want variety or because you are using chunky ingredients, go ahead and do so before capping.

5. **Cap your bottles:** Make sure the caps are on tight, but not too tight! As the pressure in your bottles increases during secondary fermentation, the caps will get a little tighter so don't crank them down. Tighten them just enough to create a seal.

6. **Secondary fermentation:** Set your bottles out of the way in a cool but unrefrigerated location. If you like your 'buch on the sweet and still side (no bubbles), you can move on to Step 7.

7. **Stop secondary fermentation:** When you are ready to halt the secondary fermentation process and get to drinking your brew (discussion below), go ahead and put your bottles in the fridge. This will slow down the activity of the microbes to a snail's pace and hold your brew at a refreshing drinking temperature.

Additional Information About Bottling and Secondary Fermentation

Now that you know the bottling drill, let's talk about what's going on in those bottles so that you can make informed decisions about how to flavor and when to drop them into the fridge.

Secondary fermentation is an extension of primary fermentation. There are residual nutrients in the kombucha to keep those microbes working away. If you added flavoring, there may be new nutrients available to give the microbes a fresh burst of energy. The same microbial population is still shacking up in there. So what's the diff?

The main distinction between primary and secondary fermentation of kombucha is that what is created in the closed bottle stays in the bottle. So as the yeasts gobble up those nutrients and turn them into CO_2, that CO_2 stays put until you pop the cap and are seduced by the hiss and mesmerized by the prick of tiny little bubbles on your tongue.

> **Yeasts are the primary CO_2 makers in kombucha. Yeasts are also the primary alcohol makers in kombucha. Secondary fermentation is when you really let the yeasts do their business and you trap it all in, alcohol included. If limiting or encouraging alcohol levels in your kombucha interests you, skip down to the section in this chapter titled "Alcohol in Kombucha."**

The ideal temperature range for secondary fermentation is between 55 and 65°F. In this range the microbes slow down, and we like to think of them as putting more thought toward what they are making. The result is a more balanced flavor profile that leans away from the strong vinegary flavors. Think about the flavor profile of lager beers—they are smooth and highly drinkable beverages. It is a similar cool fermentation process that leads to lagers, and this strategy also works very well for kombucha. Higher temperatures during secondary fermentation can be volatile and lead to bitter and super-tart acetic acid characteristics.

Secondary fermentation can last anywhere from two days to two years, depending on what you are going for. The variables are how sweet your 'buch was when you bottled it, what flavoring components were added, if any, and at what temperature your secondary fermentation is happening. Knowing when your bottles have reached perfection in terms of drinkability is something you will learn through trial and error with your unique environment and ingredients. A good reference point is two to four days at cool room temperature with sugary flavorings, and seven to fourteen days with no sugary additions. In 'Buch Kamp III we talk more about flavorings and how they react during secondary fermentation.

If you have the patience for a long, slow secondary ferment, the delicious results will be worth the wait. Secondary fermentations of this magnitude would happen in close to refrigeration temperatures, like in a wine cooler or root cellar. Around 47°F is where you would want to keep your bottles for a long-haul fermentation. Test one every few months to see where they are along their path to awesomeness.

A batch of 'buch that went too far in primary fermentation can be your secret weapon in secondary fermentation. If you have 'buch that is overly acidic coming out of primary fermentation, brew up a separate batch of tea that has dense fruity or floral flavors, like hibiscus or mandarin orange. Sweeten that new tea to taste as if you were going to drink it on its own. Blend that with your overfermented 'buch and bottle the blend. A good recipe is three-quarters 'buch to one-quarter sweetened tea. The extra nutrients will give your 'buch the fuel it needs to make awesome bubbles, the tea combination will create amazing, complex flavors, and your final yield will be 25 percent more! It's nothing but net.

It is common for first-time kombrewers to not reach their desired level of bubbles after secondary fermentation. Don't fret if you don't get it the first time.

Even when you become more comfortable with the process, there will be the occasional bottle that just doesn't fizz. Remember the process in which CO_2 is formed and you will be fine. An abundance of bubbles requires happy yeast and food for that yeast (aka sugar), so if your bottles are lacking in the bubble department, they are probably lacking in one of those two. Also check the caps to make sure they aren't leaking gases.

A tried-and-true technique for generating bubbles without changing the flavor profile of your Straight-Up 'Buch is a method we like to call the German Grandmother. Add two or three raisins to your bottles before filling them with 'buch. The extra sugars and yeast coating the raisins will give your brew the ingredients it needs to make nice soft bubbles.

In the pursuit of beloved bubbles, you must be hyperaware that you are creating a high-pressure situation within your bottles during secondary fermentation. The bottles we laid out for you at the beginning of this chapter can withstand some pressure, but there is always the possibility that they will crack or even explode. To be safe, place your bottles in a closed cardboard box or a closed cabinet during secondary fermentation to prevent exploding glass from hurting you or someone you love.

Once you are satisfied with the duration of your secondary fermentation, move your bottles to the refrigerator or root cellar. As we mentioned before, microbes have ideal temperature ranges within which they are active and kombucha microbes slow way down in refrigeration-style temperatures. If they are left in there for many months, you may notice the flavor profile starting to further change as they continue to slowly ferment. If you leave them in there for more than a month, you may also start to see your bubble content go down. But with your style and technique, you are creating an irresistible brew; we seriously doubt they will sit there for that long.

Some flavoring ingredients have just the right balance of nutrients and sugars such that refrigeration doesn't stymie secondary fermentation like you think it would. Strawberries and ginger together sometimes do this. When you find a flavor that secondary ferments so fast that you can't control it and it winds up going straight to vinegar, next time you bottle it, put it straight into the fridge. It may take a little longer to achieve your perfect level of bubbliness, but it will be *sooo* good.

The daily 'buch drinker will have a steady stream of new batches coming down the pipes at all times. There will always be 'buch in primary fermentation, 'buch bubbling away happily in secondary fermentation, and 'buch chillaxin' in the fridge. Confusion can set in real quick if there isn't a proper labeling system in place. Use sticky labels or masking tape to label your bottles with the ingredients, bottling date, and the date you intend to switch it to refrigeration. The more info, the better.

Alcohol in Kombucha

Natural fermentation processes involving yeast and carbohydrates result in the production of small amounts of alcohol. Ferments that contain natural yeast, like kefir, sourdough cultures, and kombucha, all contain small amounts of alcohol even though they are considered nonalcoholic products. In the case of sourdough, the baking process evaporates the alcohol to make it a non-issue.

With raw kombucha, the amount of alcohol in your bottles will depend on several factors. When kombucha comes out of primary fermentation in the ferm vessel, it will usually be somewhere around 0.2 to 1% ABV (alcohol by volume). To give you an idea of how kombucha stacks up right after primary fermentation, here are some average ABV values of a few common beverages:

Soft drinks—0.3%
Nonalcoholic beer—0.5%

Lager beer—4%
Dark beer—6%
White wine—9%
Red wine—14%
Liqueur—22%
Hard liquor—44%

Kombucha just out of primary fermentation will typically contain about the same amount of alcohol as a soft drink or nonalcoholic beer. Alcohol levels stay in check while in the ferm vessel because of the activities of Acetobacter and natural evaporation. The alcohol checks and balances system in kombucha looks something like this:

1. Yeast converts sugar to alcohol.
2a. Some alcohol evaporates from the surface and escapes through the cloth cover.
2b. Some alcohol is consumed by kombucha bacterial species like Acetobacter, which uses it to make acetic acid.
2c. The small amount of alcohol that remains makes up the small ABV in your home brew.

Because of these checks and balances, kombucha in an open vessel will maintain that low alcohol level and will be hard-pressed to reach higher levels. Although kombucha is naturally a low alcohol ferment, with a little manipulation, your home brew can reach higher levels of alcohol if that's your goal.

During secondary fermentation, the bacterial population of your kombucha is greatly affected by the new environment that is quickly established once you seal that bottle. Bacterial functions are greatly diminished if not stifled altogether without access to their sweet nectar of life: oxygen. Acetobacter in particular requires oxygen to convert alcohol to acetic acid. When this happens, the yeasts' ability to convert sugars into alcohol go largely unchecked and significant increases in ABV can occur. A sweetened bottle of 'buch can get up to 2% or even as high as 3% ABV during secondary fermentation if left to its own devices.

If you are an adult home brewer who wants to bring their 'buch to a Dionysian soirée, check out the section at the end of 'Buch Kamp III on alcoholic 'buch recipes.

If you are looking for ways to control your alcohol content, here are a few suggestions for keeping the alcohol on the down-low.

- Put your bottled 'buch into refrigeration immediately after bottling.
- Strain off as much yeast as you can before bottling your 'buch. You may even want to invest in a more intense yeast filter such as a cartridge filter.
- Flavor your 'buch with foodstuffs and teas that won't carry additional sweeteners.

Now . . . pour yourself a big cold glass and get ready to take your 'buchy love affair to the next level. Flip on to 'Buch Kamp III and prepare to become a kombrewing legend.

'BUCH KAMP III: A HIGHER LEVEL OF TASTE AND KNOWLEDGE

You've learned how to turn sweet tea into a radically healthy and delicious probiotic beverage. You're now a kombrewer. Awesome job. While this is quite an achievement, when you reach a goal you should take a moment to revel in your victory . . . and then set the bar higher. There is always a next step and a multitude of paths to get us there. The skills laid out in 'Buch Kamp III are born of our successes and failures. It must be said that flavoring kombucha is a skill that will come with experience. The more you learn, the more you will be free to let go and try new things.

One thing we always did on our journey to becoming kombrewmasters was never hold back. If we conceived it, we did it. Nothing was too outlandish and nothing was off-limits. This simple philosophy has allowed us to create some of the best expressions of this wonderful ferment. As we continue down our paths as kombrewmasters, we keep brewing, keep discovering. Always.

The philosophy that we put toward trying new things in kombucha brewing is the same philosophy we use in our business, in our marriage, and in the rest of our lives. Some failures will happen, sure. But whatever. We learn, we get better, find awesome new tricks that we would have never stumbled upon. Be bold. Try things. Carpe diem.

Flavoring Considerations

When flavoring your 'buch, the taste of your finished product is definitely the best place to start because at the end of the day, kombucha is already healthy . . . now let's make it delectable. But there are some other things you might consider while choosing flavoring ingredients.

How much sugar will it provide for my yeast during secondary fermentation?

As you learned in 'Buch Kamp II, the amount of sugar you add to your bottles during secondary fermentation will in part determine how bubbly your 'buch gets and how fast it gets there. We will give you a rough idea of how each of the flavoring ingredients listed below will act in secondary fermentation.

Can I increase the superfoodiness of my 'buch by adding other superfoods?

If you have a particular superfood that you love to eat to increase your overall vitality, why not try it in your 'buch? Chia seeds, turmeric, goji berries, açaí berries . . . we've tried them all, and in such dynamic partnership, teaming up definitely wins the prize.

What's in season?

It is ancient wisdom that your overall health and vitality will improve if you eat what is in season in your locale. You are a part of the synergy and overall unanalyzable essence of wholeness of your ecosystem just as much as a watermelon in the summer or an orange in the winter. Be one with your home by eating what is available close by.

Is it something that I add during primary or secondary fermentation?

There are as many flavoring options as there are microbes in a one-gallon batch of 'buch. But not all flavoring options are created equal. Many of our most beloved fruits, roots, herbs, and spices have antimicrobial properties. Although that can be of benefit to your health, it is obviously not so great for our microbial brethren. For a consistently robust home brew, carry out primary fermentation with only basic green, black, and/or white teas. Leave the additional flavoring to secondary fermentation after you've removed the SCOBY and the majority of the culture's activity is complete.

Now that you have the gist of it, flavoring can be broken down into two categories: tea blending and adding flavoring foodstuffs.

Tea Blending (Primary and Secondary Fermentation)

As we said at the beginning of this chapter, the art of tea is an extensive subject. We will hardly even scratch the surface in this book. If your passion is ignited, it is highly recommended that you get more intimate with the subject and become more connected with the process. There are hundreds of books on the subject of tea as well as websites and documentary movies. Many teahouses offer educational programs, and we do too in the KBBK Learning Center. Even with very basic knowledge, you can still create interesting blends that will give your SCOBY all the nutrients it needs while creating complex and beautiful flavors for the drinker to behold.

Here are some of our favorite recipes to get you going. When you find something that really sings, post on one of our social media outlets! We'd love to try it too.

Super SCOBY Dragon Blend—Primary fermentation: 1:1 oolong and Darjeeling. Secondary fermentation: 3:1 Oolong/Darjeeling 'Buch and White Peony tea sweetened to taste. Age 14 to 21 days at cool room temperature.

> Cool room temperature is in the sixties or low seventies.

Red Giant Blend—Primary fermentation: English breakfast. Secondary fermentation: 1:1 English Breakfast 'Buch and hibiscus sepal tea sweetened to taste. Age 14 to 21 days at cool room temperature.

Flower Power Blend—Primary fermentation: jasmine green. Secondary fermentation: 3:1 Jasmine Green 'Buch and equal parts rose and lavender tea sweetened to taste. Age 14 to 21 days at cool room temperature.

Experiment freely with sweeteners in secondary fermentation. Different sweeteners lend different flavor characteristics to your kombucha, and during secondary fermentation there's no risk in changing the microbial culture you are propagating.

Chai Spice Up—Primary fermentation: English breakfast. Secondary fermentation: 1:1 English Breakfast 'Buch and chai tea sweetened to taste with honey. Age 14 to 21 days at cool room temperature.

Adding Flavoring Foodstuffs (Secondary Fermentation Only)

The perfect recipe of flavoring components for your kombucha can be as simple as mixing in some of your favorite juice or fresh fruit, or it can be as complex as combining many different ingredients to create the perfect 'buch symphony. Below are a few of our favorite combinations. Any amount of most of these will be an awesome addition to your brew. Clearly you wouldn't want to drink a 50/50 blend with straight ginger juice! But a 50/50 blend with peach juice is sublime. And so is a 90/10 blend. Don't be afraid; kombucha is a really forgiving base. Use the list to kick-start your imagination and try different amounts.

The Simple Additions

Açaí berry—Going tropical is always a good idea when flavoring your 'buch. Açaí berries are a good choice for two reasons. First, they sweeten up your 'buch in the most pleasant way. Second, açaí will affect a beautiful deep hue of purple throughout your home brew. Never underestimate the power of aesthetic beauty in food. Juice and frozen purees from the health food store both work well. Açaí has natural oils that could cause it to separate from your 'buch. Swirl well before drinking.

Blackberry—Another easy crowd-pleaser. Blackberries are packed with flavor and have a dark reddish hue that creates a beautiful bottle of 'buch. Go with juice or crushed fresh berries on this one.

Blueberry—Everyone loves this ultrasweet blue mountain fruit. When paired with 'buch, it doesn't disappoint. Crushed fresh berries, juice, or super-freeze-dried or dehydrated fruit will be bluetiful in your brew.

Cherry—Dark cherries will add sweetness and tart cherries will add tartness. As obvious as that is, cherries of any variety are instant bedfellows of your brew. Crushed fresh, dehydrated, super-freeze-dried, and bottled cherry juice all do very well.

Concord grape—This dark delight will give your brew a gorgeous sweetness and delectable flavor. A great flavor choice for kids. Even kombucha haters love drinking this one. Fresh-juiced fruit, from concentrate, and freeze-dried all work well. We use Concord grape in KBBK's Grape Equalizer.

Ginger—To be fair, ginger deserves an entire section of its own. It probably has the most complementary flavor profile for your 'buch, and it comes with the most possibilities for combinations. Mixing spice with a hint of sweetness, fresh-pressed ginger juice will give you the most booyah, but pieces of sliced fresh ginger will also do the trick.

If you don't have a juicer, use a Microplane to finely grate the ginger. Take the gratings, wrap them in cheesecloth, and *squeeeeze* out the juice!

Grapefruit—You either love it or hate it. Grapefruits are the citrus family's bitter member. When they mingle with kombucha, it's a dry, grown-up party. If that's not your thing, pink grapefruits are a sweet reprieve. Fresh juice with a pinch of sea salt will elevate your brew.

Hops—The beer drinker's 'buch! This is one of the finest ways to pay homage to beer making without the hassle of actually making beer. Hops are a delicious floral addition to 'buch. Our favorite method is to use whole dried hops. Cascade, Willamette, and Chinook are a good starting place in your trials. For the experienced hophead, try using an extract made from whole hops or hop pellets. All forms are readily available at your local home brew shop or online.

Lavender—Everyone needs a break. Everyone needs some comfort. For those of you looking for a place to unwind and think pure thoughts, this brew is for you. Purple lavender flowers give an almost grapelike finish. Flower petals are best, but the extract will work too.

Mandarin orange—Citrus is your 'buch's friend, and mandarin oranges are your 'buch's BFF. They lend an extra bit of sweetness and bring up the rear with a wonderful florality that only citrus can impart. Use fresh-squeezed juice for a sweet and floral 'buch. For floral flavor that is on the dry side, use the rind.

Peach—This southern belle will have quite an effect on your home brew. The light acidity and the peach floral sweetness transform kombucha into a mouthwateringly juicy wonder. Crushed fresh fruit, puree, dehydrated, and juice will all work well. Try adding a dash of cinnamon to make a peach cobbler 'buch!

Pear—Simple. Sweet. 'Nuff said. We love to add pear when the weather starts to turn cold and we need an easy drinking comfort 'buch to prepare us for the long Northeast winters. Pear cider, pear juice, or pear puree will make it all better.

Raspberry—Balanced on the line between sweet and sour, raspberry bounds in with a kick! 'Buch haters will become 'buchheads under the power of the razzmatazz. Puree, crushed berries, dehydrated fruit, or juice; whichever you use, try adding a few pieces of ginger to spice things up. We call that a Razzle Dazzle.

The Complex Additions

Apple mint—Apple juice or apple cider along with mint yields a sweet, green, refreshing brew that is easy to drink and warming to the spirit. Add several crushed mint leaves per bottle and enough apple to taste.

Beet carrot ginger—This is hands down one of our favorite 'buch blends. Not just because it tastes wonderful but also because it gives your body an extra dose of hell yeah! Fresh juice for all three is recommended here, equal parts beet and carrot with ginger to taste. It's known in these parts as the Hangover Killa.

Chamomile lemon—Take note: Adding acidity not only intensifies tartness, but also can coax out the innate kombucha florality. The delicate yet robust combo of mild chamomile and lip-smacking lemon make for an almost out-of-body experience. The chamomile calms the lemon, but only after its citrus sour pulls out the innate attributes from within the 'buch. Whole chamomile flower or strong cooled chamomile tea and fresh lemon juice will bring this flavor home.

Chile pepper lime salt—If a party is what you're after then this is the flavor for you. We call it the Kombuchalata. A house favorite around KBBK, the salt's minerality and the spicy finish of the pepper make this detoxifying drink a must-try. On the easy, use Tabasco for heat, fresh lime juice to taste, and a pinch of sea salt.

Lemon ginger—Floral and fruity added to spice, oh yeah. Squeezing in fresh lemon juice is the easiest and tastiest method of lemony deliverance.

Lemon mint—We call this La Niña. A perfect goddess wrapped up in a perfect storm. Yellow tart and green spice. Fresh lemon juice to taste and a few fresh crushed mint leaves.

Mango elderberry—Mangoberry. As sophisticated as it sounds, this is both easy to make and easy to drink. You will get a burst of tart tropical fruit and smooth berry sweetness out of it. Along with the acidity from your 'buch, this tropical blend will be a crowd favorite and may prevent colds! Juice, extracts, and/or dehydrated fruit will work best.

Pineapple coconut lime—Lime 'n da coconut! If you go here, you're not far from a delicious kocktail—adding rum to this blend will take you all the way to Rio. Fresh and bottled juices alongside dehydrated fruit chunks work well. Coconut products typically have natural oils that will cause it to separate from your 'buch. Swirl well before drinking.

Root melody—A step in a more complex direction, but the reward makes the journey so worth it. Birch, burdock, sarsaparilla, sassafras, and a touch of vanilla will make your brew taste more like root beer than kombucha. Make an extract on the stovetop with fresh roots and sweetened water to yield an intense flavor addition. Alternatively, use extracts. Both will leave you smiling.

Rose and orange bitter—Working with floral flavors brings out the latent flower power in 'buch. And the added bonus: The additional floral compounds mute the funky notes that sometimes are expressed in kombucha, leaving nothing but beautiful fruit. The combo of rose and orange bitter yields a dual experience. First you get the aromatic qualities on the nose, then it shifts to magnum opus as it crosses over the palate. The rose is for the nose and the orange tantalizes the tongue. Rosewater or whole rose flowers and bitter orange extract or orange peel in any combo will impress the pants off the one you love.

Strawberry ginger—Sweet red berries to balance that fiery heat. Chopped fresh, juiced, dehydrated, super-freeze-dried, and pureed strawberries all work exceptionally well.

Watermellow twist—Did we hear someone say "summertime all the time"? Well, we did! This watermelon and lime kombo is a skinny dip for your 'buch. Fresh juice all around yields an out of this world blend. Blend both to taste and don't skimp on the watermelon juice! This is the recipe for our KBBK flavor of the same name.

Alcoholic Kombucha Production

If you like to get your booze on, this section is for you. Many turn away with disappointment when they ask the alcohol level of 'buch. Don't go! It's very easy to re-ferment your 'buch and get those alcohol levels higher, while at the same time reaping the exceptional health benefits of kombucha.

To do this, it will require that you gather more hardware. In fact, you will be setting up either a small-scale wine or beer operation. The idea is to take your kombucha straight out of primary fermentation and pitch, or add, additional hyperactive yeast. You will then put this mixture in a sealed container and continue to ferment it.

By using yeasts specific to wine or beer, you will get some of the more subtle flavors of those ferments. You can also add ingredients to this fermentation stage and enhance the final flavor. Here is one recipe to get you started. We love to brew this for our office parties at KBBK.

'BUCHAWINE

Mix one gallon of 'buch with a solid dose of pomegranate juice (about 1 cup) and one standard-size package of champagne yeast. Allow the mixture to ferment at room temperature for 7 days in a glass fermentation vessel with an airlock. Glass growlers work great for this; you can pick up an airlock at any brew supply shop or online. After seven days, bottle the mixture and put it in the fridge. The result is a tart and delicious buzz at around 6% ABV.

Graduation

Your journey through 'Buch Kamp has now come to a close. We have gone over all of the necessary steps and ingredients to make the best damn kombucha on the block. Like your SCOBY, pass your knowledge along to friends and family. Be adventurous with your brew. Try new things. Look outside these pages for inspiration and discover kreative kombos for the next generation of 'buch brigadiers.

Don't let this be the end of your fermentation journey. Now that the jar has been opened and you have experienced the magic of fermentation, keep going! Cut up some cabbage, salt it, and jam it into a jar: Make sauerkraut! Find a starter, mix it with milk, and make yogurt or kefir or villi. Try your hand at some fresh chèvre. Or perhaps brew up some English pale ale! The options are almost endless.

Without further ado, let me introduce you to the next wave of the kombucha lifestyle. It's time to take the 'buch out of the bottle and enliven your food. Let's get kooking!

DINING ON KOMBUCHA AND SCOBYS

So you're a full-on 'buch brigadier and the kombucha is flowing. Your friends love your home brew. Your mom keeps sneaking into your apartment and stealing your stash. Even your dog can't get enough of your pure liquid love. You are brewing constantly to keep up with demand, and it's forced you to become efficient and wise in your practice. There's only one problem, one loose end, one more step to full commitment.

Question: What do I do with all of the freakin' SCOBYs?

Answer: There are many awesome things you can do with those 'buch babies:

- Patch your shoes with SCOBY leather.
- Use them in a Halloween touch-and-feel game.

or

- EAT YOUR SCOBYs!

WTF?!?! Are you serious?

Don't worry; we will hold your hand. This chapter is full of easy, healthy, and delicious recipes to ease your transition from kombucha drinker to kombucha eater. If you're on the wimpy side, start with a condiment like the Spicy 'Buch Mustard. That recipe uses kombucha, or what we call kombucha vinegar (overfermented 'buch). It'll be like training wheels. By the time you've finished the jar, you'll be ready to rip into a SCOBY with your bare teeth!

With every new batch of 'buch you ferment, you'll have a new super tasty, super probiotic SCOBY. Unbeknownst to most, that little baby and the liquid it bathes in are versatile kitchen resources that rival the best gourmet ingredients. With taste and texture fit for a king, kombucha, kombucha vinegar, and SCOBYs will bring life to any dish! Quite literally.

SCOBYs are like the high-rises of microbial living, complete with elevators and unique areas where different things take place. SCOBYs are highly organized, having distinct layers of microbial species and channels that gases and liquids are transported along. The SCOBY is constructed from cellulose, the same substance that makes up the cell walls of green plants and algae. Cellulose is dietary fiber. So when you eat a SCOBY, you are eating probiotics, some of the liquid 'buch and cellulose—super-yummy, super-healthy.

Raw kombucha and SCOBYs are as live as can be. If you heat them above about 118°F, you are essentially pasteurizing them. At that temperature, biological molecules like proteins and vitamins start to change—bonds are broken, molecular structures are damaged, health-giving properties are attenuated. Microbes will also start to die off at that temperature and perhaps even a little before. But that's okay. You don't have to eat every ounce of kombucha raw. Sometimes we like our broccoli raw and sometimes we prefer it steamed. You get what I'm sayin'? We have included recipes for eating your 'buch both raw and cooked. We encourage both.

Now, let's get cooking with kombucha! The recipes that follow are just suggestions to get you started. As 'buch brigadiers, it is expected that you will take this practice to the next level. Use kombucha in your tried-and-true recipes for breads, smoothies, entrees, and desserts—use it where your recipe calls for liquids like fruit juices or vinegar. SCOBYs can be used to add texture to recipes that need a little something, or use them in the blender as an all-natural thickener. Use them in whatever ways you can dream up, but don't hold back! Share them with the world on one of our KBBK social media outlets. This is a kombucha kulture we are building, one foodie experiment at a time.

One night, KBBK team member Dan "the man" Schroder was camping out after selling 'buch at a music festival. Having only a grill, some hummus, a pita, and a SCOBY, Dan grilled his SCOBY, chopped it up, placed it in a nice warm pita with hummus, and voilà! The healthy SCOBY sloppy joe was born.

SCOBYs are compostable. In fact, incorporating SCOBYs into your compost will stimulate the action in there and make it decompose faster into a stronger fertilizer for your garden.

Kombucha Condiments

SCOBY Salsa Ménage

Spicy 'Buch Mustard

Kombucha Breath of Fire (Aceite Picante)

KomBlueberry Quick Pickle

Simple Kombucha Salad Dressing

Soups Alive!

Carrot-Ginger-SCOBY Soup

Super Live Miso Soup with SCOBY Noodles

Vegan Kombucha Dashi

Mrs. Childs's Chicken Sniffle Soup

Scrumptious Sides

KomQuinoa Pilaf with SCOBY and Roasted Root Vegetables

SCOBY and Red Pepper–Stuffed Acorn Squash

KomBrasied Kale with Cranberries and Almonds

AvoKombo

Kombucha Entrees

Eggs Poached with Kombucha

Kombucha Marinade for Meat and the Meatless

Kombucha Rub-n-Sauce for Pork Tenderloin and Tofu Block

Detox Dog

SCOBY Tempura Salad

Bread, Sweets, and Treats

Kombucha Apple Pie

KBBK's SCOBY Rancher's Snacks

Kombrewmaster's Smoothie

Easy 10-Grain Sourdough Kombucha Bread

SCOBY Salsa Ménage

Makes 5 cups

Salsa is traditionally a monogamous dance. But under the influence of that sexy tropical rhythm, it is known and understood that swapping partners from time to time adds a most pleasurable kick to a hot night. Fresh salsa—the condiment, not the dance—is such an enlightened mixture that when superfly kombucha shows up to the party, it's love at first sight and a perfect ménage ensues.

Feed this to your guests at your next summer party and watch how well everyone gets along. This recipe is truly delicious in its simple form, but if you want to have a thrilling night, throw in some grilled corn or pineapple and watch your party kombuchasize!

2 whole jalapeño peppers
5 tomatoes, chopped
1 onion, chopped
1 clove garlic, finely chopped
½ bunch cilantro, chopped
½ cup chopped kombucha SCOBY
½ cup chopped pineapple or raw kernels from 1 ear of corn (optional)
Salt and pepper to taste
1 lime

That cellulose patty that we call a SCOBY is strong—real strong! Try tearing a SCOBY with your hands. A healthy SCOBY will rip, but not without a good amount of strength. Even a knife sometimes has difficulty cutting through, so try using a pair of kitchen shears to cut up your SCOBY.

1. Roast the jalapeños. If it's summer, you might have a grill going; if so, plop your peppers on your hot grill and rotate until all sides are charred. If you are doing this on your stove, find a way to safely and securely hold your peppers either by skewering them or gripping them with a pair of tongs. Hold the peppers close to the burner or flame and rotate as necessary until all sides are charred. Allow the peppers to cool down, then peel off most of the charred skin, leaving behind the soft and smoky pepper flesh. Remove the seeds and finely chop the flesh.

2. Put the chopped tomatoes into a nice bowl. Once this salsa is complete, you won't have time to transfer it to your serving dish before it starts getting gobbled up—it's that awesome. Add three-quarters of all the ingredients except the jalapeños and lime. Taste. Adjust any of these ingredients as needed with the leftover one-quarter.

3. Add one-quarter of the jalapeños you've chopped. They can be sneaky, so add them slowly. Once you have settled on the right heat, add squeezes of lime juice until you acquire the perfect balance.

4. Devour with chips or soft warm tortillas.

Spicy 'Buch Mustard

Makes 1¼ cups

This recipe is good for those just dipping their toes into cooking with kombucha. If you are a little skeptical about eating your 'buch, this one is low on labor and high on reward.

Around our house we rarely make it past step 1! We start digging into the 'buch-soaked mustard seeds on day three as is. At this stage, you can throw them into your sautéed greens to spice things up or spread them out on your sandwich to add some shebang to that croque-monsieur. This simple, unadorned concoction can be used in many creative and heat-expanding ways. Follow the full recipe for a smooth, fine-ground mustard.

½ cup whole brown mustard seeds
About 1 cup kombucha vinegar
1¼ teaspoon turmeric
¼ teaspoon sea salt
½ clove garlic

1. Put the mustard seeds in a lidded small (about 9-ounce) glass jar and cover them with vinegar. You want about ¼ inch of vinegar rising above them. Put the lid on the jar loosely and let stand for at least 3 days. It can go up to 2 weeks on the shelf like this, which is great because it gives you a window of time in which you can finish making the mustard. Make sure you keep an

eye on the seeds, and if they start poking out of the top of the vinegar, add more so they stay submerged.

2. Pour the contents of the jar into a blender. If you haven't been dipping into the soaking mustard seeds all along, go ahead and add all of the turmeric, sea salt, and garlic. If you have taken some out, adjust these ingredients accordingly. It's not rocket science, so you are good just eyeballing it. Add ¼ cup more vinegar and puree the mixture, periodically scraping the sides with a spatula. Add more vinegar to obtain a smoother consistency. The longer you puree and the more vinegar you add, the less grainy your mustard will be. Serve with pretzels! Bangers! Pork chops! Sandwiches! The list is endless. . . .

Kombucha Breath of Fire (Aceite Picante)

Makes 1 quart

Every year the Brooklyn Botanic Garden hosts a Chile Pepper Fiesta. We are lucky enough to be neighbors with the garden, and this year, after everyone was finished feeling the burn, we inherited more hot peppers than we knew what to do with! When life gives you peppers . . .

We've done all sorts of spicy kombucha kitchen experiments, but none as versatile as the Kombucha Breath of Fire. When we whipped this one up, we had no idea what the potential of this concoction would be. This is by far the most used kitchen staple we have come up with to date. We dribble a little on our breakfast sandwiches, we sprinkle some drops onto a warm bowl of Chicken Sniffle Soup (page 122), and we seduce the muse by using it to dirty a martini.

1 quart spicy peppers (we like a combination of jalapeño, habanero, long thin cayenne, and large cayenne peppers)
About 1 quart kombucha vinegar

You may want to wear protective gloves while you are setting this up. It is a horrible experience to rub your eye and get an eye full of capsaicin. If you don't wear gloves, wash wash wash your hands!

1. Wash your peppers thoroughly and poke them up and down with a fork. Give one fork-poke per inch all the way down each pepper. Cram the poked peppers into a 1-quart sealable jar. Pour enough vinegar into the jar to completely cover the peppers. Use a fork to push down on the peppers and squish the air out of them. Pour more vinegar into the jar if necessary to cover the peppers. Repeat until all the air is gone.

2. Let stand, covered, for 5 days at room temperature, monitoring it to make sure the liquid is still covering the peppers. If not, top it off!

3. Pour the Kombucha Breath of Fire into another vessel, leaving the peppers behind. Store refrigerated for up to 1 month. Keep a few of the peppers in a sealed container in the fridge for up to 10 days. Finely chop them as needed and use them to add sweet heat to a variety of dishes like soups or stir-fries. *Mmm.* Hot. Yummy. Good.

KomBlueberry Quick Pickle

Makes 2 cups

As quirky as it sounds, this deliciously bumpy little treat will add flair to pretty much every conceivable dish. For a unique hamburger topping, squish up the blueberries before slapping them on. Or strain the liquid and serve it over eggs and toasted bread for a sophisticated breakfast. You can even toss into raw baby spinach with walnuts for a live superfood-rich salad!

1 cup kombucha vinegar
¼ cup water
¼ cup maple syrup
1¾ tablespoons sea salt
½ small red onion, very thinly sliced
1 pint fresh blueberries

Combine the vinegar, water, maple syrup, and salt in a bowl and mix until the maple syrup has dissolved. Add the red onion and blueberries. Mix well so that all surfaces are coated. Cover and refrigerate overnight. It will keep in the refrigerator for up to 4 days.

Simple Kombucha Salad Dressing

Makes 1¼ cups

Cooking with kombucha doesn't always have to be elaborate. Some of the most elegant dishes are the simplest ones. Salads are a concert of fresh vegetables uncomplicated with modern techniques. Simple salads require a dressing that is worthy of them. May we present to you our simple kombucha dressing that is easy enough to whip up for a solo lunch.

¾ cup extra-virgin olive oil
¼ cup kombucha vinegar
¼ cup fresh-squeezed lemon juice
1 tablespoon Dijon mustard (or Spicy 'Buch Mustard, page 111)
1 tablespoon honey
Pinch of salt
Pinch of pepper
Pinch of dried herbs

Combine all the ingredients in a bottle that you can tightly seal. Shake well. That's it. Enjoy! May we suggest a nice mixed greens salad with white beans and shredded beets? Toss in some diced SCOBY for an extraordinary texture kombo.

Carrot-Ginger-SCOBY Soup

Makes 6 (1 cup) servings

1 tablespoon plus 1 teaspoon extra-virgin olive oil
½ cup chopped red onions
2 cloves garlic, finely chopped
1½ pounds carrots, peeled and cut into large chunks
3 cups vegetable stock
½ cup raw cashews
½ cup kombucha or kombucha vinegar
1 cup chopped kombucha SCOBY
1 tablespoon fresh ginger juice
Salt and pepper

1. Heat 1 tablespoon of the olive oil in a soup pot over medium heat. Add the onions and garlic and cook, stirring occasionally, until fragrant and the onions are translucent, about 4 minutes. Toss in the carrots and cook for 2 minutes more, stirring occasionally. Add the vegetable stock and bring to a simmer.

2. Heat the remaining 1 teaspoon olive oil in a sauté pan over medium heat. Add the cashews and stir continuously until the nutty cashew smell starts to fill the room, about 3 minutes. Remove from the heat and set aside.

3. Test the carrots in the soup pot for doneness after 15 minutes of simmering. They are done when they are easy to pierce with a fork. When the carrots are

cooked, turn off the heat and add the cashews to the soup. Let cool to a comfortable eating temperature, then add the vinegar and the SCOBY pieces. Transfer to a blender and puree until smooth, adding water as necessary to thin it out.

4. Add the ginger juice and season with salt and pepper. Serve immediately or cool to room temperature, refrigerate, and serve cold. A nice raw fermented sour cream, yogurt, or almond yogurt makes a great garnish.

Super-Live Miso Soup with SCOBY Noodles

Makes 4 (1½ cup) servings

2 tablespoons dried wakame seaweed

4 cups Vegan Dashi (recipe follows), traditional dashi, or chicken broth

¼ cup sliced shiitake mushrooms

1 small leek, white part only, sliced

¼ cup red miso paste

1 tablespoon tamari

8 ounces firm tofu, at room temperature, cubed (optional)

1 cup shredded kombucha SCOBY (use the shredding plate on a food processor), at room temperature

About ¼ teaspoon toasted sesame oil

1. Place the wakame in a bowl, cover with water, and set aside for 15 minutes. Drain the wakame, chop it, and set aside.

2. In small soup pot, warm the dashi over medium heat. Add the mushrooms and leek, and simmer for 4 minutes. Turn off the heat.

3. Once the broth is cooled to a comfortable eating temperature (below 118°F), remove ¼ cup of the broth, pour it into a small bowl, and whisk in the miso

paste. Add the miso mixture back to the pot with the remaining broth and add the tamari. If you are using tofu, add it now.

4. Place ¼ cup shredded SCOBY into each of 4 bowls and pour 1¼ cups of the warm soup over them.

5. Top with a little stack of wakame and a few drops of toasted sesame oil.

Vegan Kombucha Dashi

Makes 4 cups

5-inch piece kombu seaweed
2 shiitake mushrooms
4 cups water
1½ tablespoons mirin
¼ teaspoon honey
2½ tablespoons shoyu or soy sauce
2 tablespoons kombucha

1. Place the kombu, mushrooms, and water in a soup pot. Cover and let sit for 1 hour at room temperature.

2. Bring to a simmer, and simmer for 5 minutes.

3. Strain out the kombu and mushrooms, and stir in the mirin, honey, shoyu, and kombucha. The dashi will keep, refrigerated, for up to 5 days, or freeze it until ready to use.

Mrs. Childs's Chicken Sniffle Soup

Makes 6 (1 cup) servings

Alas, there are some flus that even kombucha can't keep at bay. When flus, sinus infections, or allergies happen, flatten that beast with this health tank of a recipe. Kombucha Breath of Fire, thyme, and oregano provide your nose with all it needs to clear the passages. The rest is soothing to the throat, nose, eyes, mind, and soul. Substitute seitan and a vegan stock for a vegan soup.

3 tablespoons extra-virgin olive oil
1 small onion, sliced
1 clove garlic, sliced
1 sprig fresh thyme
1 sprig fresh oregano
Salt
2 small carrots, sliced
2 celery stalks, sliced
1 quart chicken stock
1 cup water
2 bay leaves
⅛ teaspoon black pepper
2 cups Kamut pasta spirals

¾ pound boneless chicken thighs, rinsed, patted dry, cut into ½-inch cubes, and lightly salted
1 large kale leaf, stemmed and chopped
1½ cups halved cherry tomatoes
Kombucha Breath of Fire (page 113)

1. In a soup pot, heat 2 tablespoons of the olive oil over medium heat until it begins to shimmer. Add the onion, garlic, thyme, oregano, and a pinch of salt. Sauté, stirring occasionally, for about 2 minutes, or until the onion and garlic are translucent. Add the carrots and celery and continue to sauté, stirring occasionally, for 3 minutes.

2. Add the chicken stock, water, bay leaves, and pepper. When the liquid in the pot is at a good simmer, add the Kamut pasta spirals.

3. Meanwhile, heat the remaining 1 tablespoon extra-virgin olive oil in a medium skillet over medium-high heat. Add the chicken thighs and brown on all sides, about 4 minutes.

4. Add the browned chicken, kale, and tomatoes to the soup pot and continue to simmer until both the pasta and the chicken are fully cooked, about 8 minutes.

5. Season with salt and serve in bowls with a few teaspoons of Kombucha Breath of Fire dripped on top, the amount depending on your taste for heat.

The scent of the KBOF will be so intoxicating, you will want to add more and more! Drip in more as needed. Toss in a chopped pepper left over from making your Kombucha Breath of Fire and you might as well lick the sun! Serve with a slice of Easy 10-Grain Sourdough Kombucha Bread (page 152).

KomQuinoa Pilaf with SCOBY and Roasted Root Vegetables

Makes 4 (½ cup) servings

Quinoa! Can you say it? *Keen-wah!* It's delicious. It's nutritious. It's filling lots of niches . . . in my belly! This is a recipe we came up with to satisfy our winter craving for hunkering down and roasting things. There's nothing better than the smell of veggies roasting on a cozy winter afternoon. Enter kombucha SCOBY and kombucha vinegar and you've got comfort on a plate.

1 small beet
1 medium carrot
1 tablespoon plus 2 teaspoons extra-virgin olive oil
Salt and pepper
1 cup water
½ cup red quinoa
2 tablespoons finely diced red onion
2 tablespoons minced kombucha SCOBY
2 teaspoons kombucha vinegar
3 basil leaves, finely shredded

1. Preheat the oven to 425°F.

2. Clean, peel, and dice the beet and carrot. Toss with 1 tablespoon of the olive oil and a pinch of salt and pepper. Spread the beets and carrots in a single

layer on a baking sheet and roast the vegetables for 12 to 15 minutes, turning once. Test with a fork for doneness.

3. Meanwhile, in a small saucepan, bring 1 cup water to a boil with a pinch of salt. Add the quinoa, reduce the heat to a simmer, cover, and cook until all the water is absorbed, 15 to 20 minutes. Toss the quinoa with a fork a few times and allow it to cool to a temperature that you can comfortably eat before proceeding. Transfer to a serving bowl.

4. With a fork, toss the roasted vegetables, onion, SCOBY, and kombucha vinegar into the quinoa.

5. Add the basil just before serving, either tossed in or as a garnish. Serve immediately, or cover, refrigerate, and serve chilled.

SCOBY and Red Pepper–Stuffed Acorn Squash

Makes 4 (1½ cup) servings

2 acorn squash, cut in half, seeds removed
1 tablespoon extra-virgin olive oil
2 cups ¾-inch bread cubes
½ cup diced red bell pepper
½ cup diced green onion
½ cup diced kombucha SCOBY
1 teaspoon dried oregano
½ cup chicken or vegetable stock
1 egg (or use flax alternative; see page 128)
1 tablespoon maple syrup
½ teaspoon salt
Pinch of pepper
2 tablespoons chopped fresh parsley

1. Preheat the oven to 350°F.

2. Brush the tops of the squash (where the cut was made) with olive oil and place faceup on a baking sheet covered with parchment paper or aluminum foil.

3. In a large bowl, toss together the bread cubes, bell pepper, green onion, SCOBY, and oregano.

4. In a separate bowl, whisk together the stock, egg, maple syrup, salt, and pepper.

5. While tossing the bread mixture with tongs, slowly pour in the liquid mixture. Using a spoon, stuff the acorn squash halves with equal amounts of filling.

6. Place in the oven and bake for 35 to 40 minutes. Look for a nice browning of the filling and test the doneness of the squash using a fork. Serve topped with the parsley.

Well-nourished and happy chickens lay eggs of outstanding nutritional value. But flaxseeds are also incredibly nutritious. Change things up by using flaxseeds in place of eggs in your recipes. It's really easy to do, and the flaxseed alternative really does make a great substitute. Buy your flaxseeds whole and grind them in a nut or clean coffee grinder. One egg = 1 tablespoon of ground flaxseed + 3 tablespoons of water.

KomBraised Kale with Cranberries and Almonds

Makes 4 (¾ cup) servings

Kale is a superfood. Cranberries are, too. Kombucha's in that mix, and . . . hey, wait. Aren't almonds a superfood as well? That's right. This polychromatic side dish will quickly become a weekly offering in your kitchen with its sweet, tangy, high superfoodiness. Make extra so you can toss some cold into your salad for lunch the next day.

2 tablespoons extra-virgin olive oil
1 small onion, diced
2 cloves garlic, finely chopped
1¼ cups chicken or vegetable stock
2 teaspoons brown or yellow mustard seeds
2 teaspoons honey
½ bunch kale, stems removed and roughly chopped
¼ cup dried cranberries
¼ cup raw almonds, sliced
¼ cup kombucha vinegar or kombucha
Sea salt and pepper

This is a great place to dig into that jar of mustard seeds soaking in kombucha vinegar for the Spicy 'Buch Mustard (page 111). Those seeds will be bursting with flavor and a little softer than if you were to toss them into this recipe dry.

1. Heat the olive oil in a large saucepan over medium heat. Add the onion and cook until translucent, about 4 minutes, stirring occasionally. Add the garlic and cook, stirring occasionally, for an additional minute.

2. Stir in the stock followed by the mustard seeds and honey. Bring to a simmer, add the kale, and cover. Reduce the heat to low and cook until the kale has softened, about 15 minutes.

3. Just before serving, add the cranberries and almonds. Add the vinegar, season with salt and pepper, and serve.

Adding the raw almonds, kombucha, and dried cranberries right before serving will help them retain higher amounts of their live nutritional value.

AvoKombo

Makes 4 (¼ cup) servings

When we're looking to be rich but still eat healthy, we look to avocados. We also look to these delicious green fruits for their detoxifying properties and as a creamy dairy alternative. (Avocado instead of mayonnaise or cheese on your sandwich is *soooo* creamily rewarding! And replacing your cheese with avocado on an egg sandwich is to die for.) This simple salad is great on its own or as a topper for crackers or a green salad.

1 ripe avocado, peeled, pitted, and cut into ¼-inch chunks
1 spring onion, diced
¼ cup kombucha vinegar
½ teaspoon lime juice
1 tablespoon toasted black sesame seeds
Pinch of sea salt and pepper

1. Place the avocado and spring onion in a medium glass bowl. Add the vinegar and lime juice, toss, cover, and let sit in the fridge for 15 minutes.

2. Remove from the fridge and sprinkle on the sesame seeds. Season with salt and pepper. Mix gently and enjoy. Chips, anyone?

Eggs Poached with Kombucha

Makes 4 poached eggs

Poached eggs are elusive to many. Eggs that fall apart or turn out like rubber pucks on your Benedict just aren't sexy. Let us give you a tried-and-true recipe using your new hot roommate—kombucha! Follow our directions, apply kombucha, and create eggy perfection every time. The secret is the freshness of the eggs combined with the organic acids in 'buch vinegar, but don't tell. Let them think it's all you.

The eggs you use to poach simply have to be fresh. You want to see three main parts to your raw cracked egg: yolk, firm egg white, runny egg white. The firmer the egg matter you have, the fresher the egg and the better the poaching will go. Let's proceed.

6 cups water
2 tablespoons strong kombucha vinegar
4 fresh farm eggs

1. In a bowl, combine the water and vinegar. Mix well and pour the mixture into a nonstick skillet until there is an inch of liquid in your pan. If you don't have a nonstick pan, use a wide and deep saucepan and bring the water level up to 3 inches, using a proportional amount of kombucha vinegar as described above. Bring to a boil over high heat.

2. Gently crack the eggs into small bowls, 1 egg per bowl. You can do 4 eggs at a time in a 9-inch skillet.

3. Gently turn the 4 eggs into the water, turn off the heat, and cover. For barely runny yolks, let stand for about 7 minutes, depending on the size, then remove with a slotted spoon. Voilà! You should be on *The Next Hot Egg Chef,* a show we just created in your honor.

Use 'buch to boil eggs too. A tablespoon of 'buch in your egg-boiling water will keep the eggs that crack from leaking into the pot. *Mmmm.* More egg for me to munch on!

When Easter comes around, don't waste your money or your health on those weird Easter egg–dyeing tablets. Food coloring substitutes can easily be made from boiling water with a coloring agent such as turmeric for yellow or beets for red (there are several websites that can show you how to do this). Once you've made your food coloring, follow this recipe for each color: 2 tablespoons kombucha vinegar, ½ cup boiling water, and up to 20 drops of food coloring substitute, depending on the depth of hue you are going for. Dip your eggs for a few minutes—longer for deeper shades.

Kombucha Marinade for Meat and the Meatless

Makes 2¼ cups

The centerpiece of most great feasts begins with a great marinade. Using kombucha will give you amazing tart undertones while tenderizing whatever it is you are marinating. The rest is pure and simple flavor, just the way we like it. Use this for all sorts of foods, from meats to meatless proteins, to veggies—anything you can roast, grill, or pan-sear.

If you are of the carnivorous tribe, think London broil or skirt steak for this marinade. If the cut you choose is more sinewy and tough, extend the marinating time and consider tenderizing mechanically as well.

If you are looking for a vegan alternative, this marinade will transform a fresh batch of tempeh or a nice firm tofu into the object of culinary desire. Cut the tempeh or tofu into the size it will be served, as this will allow more of the marinade to penetrate. We also recommend boiling the meatless protein in the marinade to reach its full potential.

¼ cup Worcestershire sauce
2 cups kombucha vinegar
¼ cup chopped onion
1 clove garlic, chopped
½ teaspoon salt
¼ teaspoon dried thyme

1. Mix all the ingredients together in a glass bowl.

2. Place the items to be marinated in the mixture and make sure all surfaces are coated with the liquid and most are submerged. Cover and let sit in the refrigerator for 12 to 24 hours. Go with the longer marinating time for thicker items.

3. For meatless protein, transfer the contents of the bowl to a saucepan and bring to a simmer. Cook for 25 minutes, then proceed to step 4.

4. Remove your items from the marinade and cook them. Don't toss the marinade itself in, but do use it as a baste.

Kombucha Rub-n-Sauce for Pork Tenderloin or Tofu Block

Serves 2

A blank canvas is such a gift in the kombuchasseur's kitchen! Without having to contend with dank overtones, floral baselines, or a pungency that just won't quit, pork tenderloin and tofu both offer the ambitious kook a virtual clean slate on which to strut his or her stuff.

Take this recipe, for example. It's a symphony of flavors that are upheld by the basic toothsomeness of the individual protein source. Like the foundation of an ancient temple, it is meant to be embellished.

RUB

2 tablespoons herbes de Provence
1 tablespoon kombucha vinegar

OTHER INGREDIENTS

1 whole pork tenderloin or 1 (14-ounce) block firm tofu, cut into ½-inch slices
Salt and pepper
¼ cup fig, peach, or plum preserves
¼ cup kombucha
2 teaspoons garbanzo bean flour (all-purpose flour is a fine substitute)
1 sprig fresh parsley

1. Preheat the oven to 425°F.

2. Mix together the rub ingredients until they form a paste. Set aside.

3. Rinse and pat dry the pork or tofu with a clean cloth or paper towel. Season well with salt and pepper.

4. Rub the protein with the herb paste, making sure to cover every surface. Place the protein on a parchment-lined baking pan, leaving space around every piece if you are using tofu.

5. Put the pan in the middle of the oven and roast. If cooking tofu, flip each piece after 5 minutes. The tofu will be finished in about 10 minutes. For pork tenderloin, roast for 12 to 15 minutes, depending on the thickness, and there is no need to flip it. Use an instant-read thermometer to determine doneness, when it reaches 145°F. Remove the pan from the oven and let the protein rest for 5 minutes.

6. While the protein rests, combine the preserves and kombucha in a small saucepan and heat over medium heat, stirring, for about 2 minutes.

7. Sift the flour into the pan and stir well to thicken the sauce. Serve the protein with the sauce drizzled over it, garnished with the parsley.

Detox Dog

Serves 4

Last year Kombuchman and friend Alexei Taylor entered to compete in Brooklyn's Great Hot Dog Cookoff, an annual competition that pits chefs against one another to win the prize for the best hot dog. Every year the winning dog is rich, decadent, and full of fat. Having the assigned starting block of a beef hot dog, 'buchman and Lex decided to approach the sport from a different angle. The team came up with the Detox Dog: a lean and mean dog with layers of carrots and kombucha-pickled daikon wrapped in a tidy nori package with a hint of creamy spiciness. It took third place out of twenty-four against some of New York City's finest chefs.

INGREDIENTS FOR THE DAY BEFORE

1 cup julienned daikon
1 cup kombucha vinegar

INGREDIENTS FOR THE DAY OF

½ cup mayonnaise
2 teaspoons wasabi powder
¼ cup plum sauce
¼ cup kombucha vinegar
¼ cup diced kombucha SCOBY
4 hot dogs or tofu dogs
8 nori sheets

½ cup julienned carrots
1 cup julienned daikon

THE DAY BEFORE:

Combine the daikon and vinegar in a sealable container. Mix well, making sure every piece of daikon is submerged. Seal the container and store at room temperature overnight.

THE DAY OF:

1. In a small bowl, mix the mayo and wasabi powder thoroughly. Set aside.

2. In a separate small bowl, mix the plum sauce and vinegar thoroughly. Set aside.

3. Strain the kombucha from the quick-pickled daikon, transfer the daikon to a bowl, and add the diced kombucha SCOBY to the daikon.

4. Cook up your dogs! We love to grill, but dogs are excellent boiled too.

5. While your dogs are cooking, toast your nori sheets by gently passing them over the grill or a stovetop burner. The texture will become crisp after just a few passes.

6. Schmear about 1 tablespoon of the mayo concoction close to one edge of the toasted nori. Slice your dogs lengthwise into quarters and place 2 quarters on the schmear. Spread approximately 1 tablespoon of the fresh carrots and 1 tablespoon of the daikon over the dog strips. Do the same with about

2 tablespoons of the kombucha-pickled daikon and SCOBY mixture. Finally, drizzle about 1 tablespoon of the plum sauce–vinegar mixture over the fixin's. Roll tightly and serve.

A bamboo sushi mat is very helpful in rolling up these Detox Dogs. Check out YouTube.com for DIY videos on sushi rolling.

SCOBY Tempura Salad

Makes 4 main course salads

1 cup unbleached cake flour
1 cup white rice flour
1½ quarts liquid coconut oil
1 large egg, beaten
1½ cups cold seltzer water
½ cup cold kombucha
1 large sweet potato, peeled and cut into ⅛-inch-thick slices
Kosher salt
¼ pound fresh broccoli, trimmed into 2 x 2½-inch florets
8 sprigs flat-leaf parsley
½ pound kombucha SCOBY, cut into ¼-inch dice
½ head lettuce, washed and chopped
½ cup Simple Kombucha Salad Dressing (page 116)

> If you don't have cake flour, you can substitute this instead: 3 tablespoons all-purpose flour sifted five times with 1 tablespoon arrowroot powder.

1. Combine the cake flour and rice flour in a medium glass bowl. Set aside.

2. Heat the coconut oil in a wok or large saucepan over high heat until it reaches 375°F. If you have a deep-fry thermometer, use it to monitor the temperature. If not, place a droplet of the prepared batter into the oil to check the temperature. You want it to fizzle and fry when you add it, but you don't want it to brown too fast. The dollop should last in there for about 45 seconds before it looks like it is browning.

3. Prepare the batter just before you plan to use it. Whisk the egg, seltzer water, and kombucha in a medium bowl. Pour the liquid mixture into the dry mixture and whisk to combine. Some lumps may remain, but try to do away with the big ones. Set the glass batter bowl inside a larger bowl filled with ice to keep it as cool as possible while you fry.

4. Dip the sweet potatoes into the batter using tongs, drain for 1 second over the bowl, then add to the hot oil. Adjust the heat to maintain the temperature as close as possible to 375°F. Fry just 6 pieces at a time so you do not reduce the temperature or overcrowd the pot. The tempura is done when it is puffy and very light golden, 1 to 2 minutes.

5. Remove the tempura pieces with a slotted spoon and place on a cooling rack. Sprinkle with salt.

6. Repeat the coating and frying process with the broccoli and parsley.

7. Finally, repeat the coating and frying process for the SCOBY pieces.

8. Arrange the lettuce on 4 plates. Place the tempura veggies around the lettuce and toss the tempura SCOBY pieces on top. Drizzle with the dressing and indulge in the healthful balance of fried on raw.

Always serve tempura ASAP. Cooked pieces may be held in a 200°F oven for up to 20 minutes, but they will start to lose their crispy texture very quickly.

Kombucha Apple Pie

Makes 1 (9-inch) pie and 1 (9-inch) piecrust to freeze for later use

This recipe is really very simple and it is a fantastic way to use your kombucha. We like to make one additional crust dough while we're making pie. It freezes well and it's handy to have pie dough in your freezer at all times. The recipe below is for one pie filling and two piecrusts. If you want to make just one crust, cut the crust recipe in half.

PASTRY CRUSTS

1½ cups whole wheat pastry flour
1½ cups unbleached white flour
3 tablespoons plus 1 teaspoon Sucanat or other minimally processed sugar
½ teaspoon ground cinnamon
½ teaspoon baking powder
¼ teaspoon sea salt
⅔ cup coconut oil, at room temperature (soft but not liquid)
2 tablespoons vanilla extract
2 cups ice-cold water
5 tablespoons Grade A maple syrup

PIE FILLING

2 cups kombucha, preferably jasmine flavored
¾ teaspoon arrowroot powder

¼ cup Sucanat, maple crystals, or other minimally processed sugar
⅛ teaspoon ground cinnamon
¾ teaspoon vanilla extract
Pinch of sea salt
3 medium apples (Granny Smiths are great for pies), peeled, cored, and sliced about ⅛ inch thick
2 tablespoons finely chopped KBBK's SCOBY Rancher's Snacks (page 149) (optional)
1 scoop vanilla ice cream (optional)

FOR THE CRUSTS:

1. Start with the crusts, as the dough needs some time to settle in the fridge. Sift the flours, sugar, cinnamon, baking powder, and salt into a large bowl. Use a whisk to makes sure the ingredients are fully incorporated.

2. Make a crater in the center of the flour mixture and add the coconut oil. Cut and blend into the flour using a spatula or pastry cutter. There should be lumps of coconut oil in the flour now.

3. Add the vanilla, 2 tablespoons ice-cold water, and 4 tablespoons of the maple syrup and mix with a wooden spoon. The dough will get clumpy. Don't mash up the clumps!

4. Add ice-cold water 1 tablespoon at a time until the dough looks consistently moist but not soaked. It will clump into many ½-inch to 2-inch balls and there should be no dry flour hanging around the bottom. Don't mash the clumps!

5. Divide the dough in half and place each half on a piece of plastic wrap. Gather up the edges of the plastic wrap and twist to make a fully enclosed ball. Flatten the balls to about 2 inches thick and place in the refrigerator from 1 hour to 48 hours. Alternatively, you can freeze the dough for up to a month.

FOR THE PIE:

1. When you are ready to make your pie, preheat the oven to 375°F and remove 1 portion of the dough from the fridge. Let the dough rest at room temperature while you prepare the filling.

2. Remove the plastic wrap from your dough ball, separate it into two pieces, one about three fifths and the other two fifths. The larger portion will be the bottom piecrust and the smaller will be the top crust. Dust the larger portion of the dough with flour. Stretch out that piece of dough with your hands a little bit, leaving it loosely symmetrical. Place between 2 pieces of parchment paper and roll out. Start each roll from the center of the dough and roll out toward the edges, keeping the dough uniformly thick and round. Occasionally lift the parchment to take a peek inside. It is okay for the edges of your dough to crack a little bit, but if cracks start occurring in the center, work a little more cold water into the area with your fingers, heal the crack, and keep going. Conversely, if the dough starts to get sticky, dust the area with more flour. When the dough is roughly 9½ inches wide, remove the top parchment and flip the dough into the pie pan. Remove the other piece of parchment and push the dough into the pan with your finger.

3. Roll out the smaller portion of dough in the same manner as the first. This one needs to be only 8½ inches around, though. Set aside until your pie is ready to be topped.

4. To prepare the filling, in a small bowl, mix ½ cup of the kombucha with the arrowroot powder until the arrowroot is completely dissolved. It will be a cloudy, clumpless liquid when it is mixed. Set aside.

5. Place the remaining 1½ cups kombucha in a medium saucepan and add the sugar, cinnamon, and vanilla. Heat until just below a simmer.

6. Add the kombucha and arrowroot powder mixture to the saucepan and stir until the cloudiness disappears, about 2 minutes. Dip a spoon in, blow it cool, and taste your tasty sauce. Adjust the seasonings to your fancy.

7. Arrange the apple slices on the bottom piecrust. Pour about 1½ cups of the thickened kombucha sauce over the apple slices—just enough to cover the apples.

8. Remove the top layer of parchment from the top layer of crust and flip it into the pie. Remove the other parchment and press the edges of the crusts together all around. You can cut off or rip off any excess crust at this point. Poke some holes all around the dough with a fork.

9. Pop the pie into the middle of the oven and let her bake there for about 30 minutes. Remove the pie, mix the final tablespoon of maple syrup with 1 tablespoon water, and brush it over your pie. Put the pie back in the oven for

5 minutes. Once your pie has a nice golden finish, remove it and set it on a cooling rack until it is cool enough to devour.

10. Serve with KBBK's SCOBY Rancher's Snacks and/or vanilla ice cream. Sit back and breathe in the fall flavor and all of your effort will melt away under the warm, smooth pleasure of kombucha apple pie.

KBBK's SCOBY Rancher's Snacks

Makes about 1 cup

This is one of our favorite uses for leftover SCOBY, one that you'll need a dehydrator to make. It is a tasty sweet treat that you can carry with you and chew on throughout the day when you need a little pick-me-up. This recipe comes from KBBK SCOBY farmer Chris Strait. He's the meticulous man who propagates the pristine and well-cared-for KBBK SCOBYs. Needless to say, his hands are often covered in SCOBY during a sort of propagation meditation. What better time to dream up new SCOBY recipes? Embodying the taste of fall, these candied SCOBYs are a tasty way to use up the cultures that pile up during a long and refreshing summer of brewing kombucha.

4 (1-inch-thick) SCOBYs
1 cinnamon stick
1 tablespoon shredded licorice root
1 teaspoon whole allspice
1 tablespoon sassafras extract
1 tablespoon sarsaparilla extract
6 cups filtered water
4 cups organic cane sugar

1. Cut your SCOBYs into small cubes and rinse them in a colander to remove tea and yeast filaments. Set aside to drain off as much as possible.

2. Combine the spices with the water until you have a thick, strong decoction (enough to submerge the SCOBYs in, about 4 cups), strain the spices out, add 3 cups of the sugar, and stir. Allow the mixture to cool.

3. Place the drained SCOBY cubes and cooled sugar water mixture in a bowl, cover, and let marinate for 24 hours in the refrigerator. Drain and refrigerate the sugar marinade for future SCOBY snack-making.

4. Toss the SCOBY pieces in the remaining 1 cup sugar, being sure to coat all surfaces.

5. Pour the sugary SCOBYs onto parchment sheets in an even layer and dehydrate at 110°F for 16 to 20 hours, until the SCOBYs are the consistency of a soft chewy leather. You can store the candies in bags or airtight containers covered in more sugar to preserve them for up to a month.

Kombrewmaster's Smoothie

Makes 3 (1 cup) servings

In order to get through the long hours in a kommercial kombrewery, a kombrewmaster needs intense fuel. This powerful concoction comes straight from the man himself: our head brewer at the KBBK kombrewery, Sam Dibble. His love and knowledge of the 'buch is unparalleled. Straight from Sam's kitchen, he is delighted to share with you the secret to his prolonged and nutritive energy, this awesome superfood smoothie.

1½ cups ginger-flavored kombucha
1 cup fresh blueberries
1 cup frozen raspberries
1 fresh banana, peeled
¼ cup hemp seeds
Leaves from 2 to 3 sprigs fresh mint

Place all the ingredients in a blender and blend on medium-high speed for 45 to 60 seconds. Pour into a glass and enjoy.

Easy 10-Grain Sourdough Kombucha Bread

Makes 1 or 2 loaves

Is there anything better than a sandwich made on a hearty sourdough slice? I'm not the only one who appreciates the marriage of flour and microbes. From Flemish desem to Ethiopian injera, pretty much every culture around the world has cultivated its own style of fermented bread using techniques similar to those used to make good ol' San Francisco Sourdough Bread: letting the flour slowly ferment in rounds with natural yeasts and bacteria.

There are some awesome bakeries that still do this labor- and time-intensive process that results in absolutely phenomenal bread. Others have "industrialized" the process by adding citric acid or another acid to the dough to re-create the sour flavor, but leaving behind the benefit of a slow ferment.

Here in our kitchen we found a middle ground by using all-natural, slow-fermented, delicious kombucha to sour our favorite easy bread recipe. The results are delish.

1½ cups white whole-wheat flour
1 cup unbleached all-purpose flour
¾ tablespoon (1 packet) granulated active dry yeast
1½ teaspoons coarse salt
2 tablespoons vital wheat gluten

1 cup uncooked 10-grain hot cereal (we use Bob's Red Mill)
1¼ cups lukewarm kombucha or kombucha vinegar
½ cup lukewarm water
Sunflower, flax, sesame, or caraway seeds for sprinkling on top

1. Sift together the flours, yeast, salt, and gluten. With a whisk, stir in the cereal. Add the lukewarm kombucha and water. Mix with a spoon or spatula until all of the dry ingredients are incorporated.

2. Cover the vessel you have been mixing in with a sheet of plastic wrap, foil, or, if your vessel has a lid, put it on there. It's not important for it to be airtight. Let the dough rest at room temperature for 2 hours.

3. You can either bake it now or put it in the fridge for up to a week and bake it later. If you refrigerate your dough, let it rest on the counter for about 90 minutes before baking to bring it to room temperature.

4. Thirty minutes before baking time, place an empty boiler pan on the lowest rung of the oven and preheat the oven to 450°F.

5. Form the dough into 1 larger loaf or 2 grapefruit-size loaves and place on a greased cookie sheet.

6. When it's time to bake, paint some water on the surface of your loaf/loaves and sprinkle on your seeds. Pour 2 cups of water in the boiler pan and slide the cookie sheet in the oven as quickly as possible. You don't want to leave the oven door open any longer than you have to.

7. Bake for 20 minutes, then remove the cookie sheet and put your loaves directly on the oven rack. Your loaves will be richly browned and firm when they are ready to come out of the oven; about 30 minutes total should do it. Let rest in the open air for a bit before slicing into your delicious and healthful kombucha bread.

KOMBUCHA KOCKTAILS

Everyone must let loose and just celebrate from time to time. Whether it is a quiet, sexy dinner or an ecstatic night of dancing to block-rocking beats, people like booze. It makes sense. We have evolved alongside boozy ferments to become the great civilization we are today!

Now that we have methods of upping the ante with higher alcohol beverages, the wise ones have figured out how to pacify the consequences. Let us introduce you to the kombucha kocktail. These boozy koncoctions follow the principle we call "reverse toxmosis"—where you detox while you retox. What better way to minimize the bad than to mix it with some good?

An easy companion to many of your favorite regulars, Straight-Up 'Buch, flavored 'buch, and 'buch vinegar are all killer mixers with killer benefits. Below is a list of our favorites. We hope at this point in your studies you are starting to get the idea: Kombucha is versatile.

Reverse Toxmosis

Serves 1

Beer isn't all bad. In fact, we are huge fans of our brother from another mother! (Get it? Kombucha mother? We know. We are dorks.) But beer comes with some less desirable qualities, such as bloating from all those carbs and aldehydes that cause hangovers. Cases of extreme laziness have been reported after swallowing down beer's hoppy goodness, and many people just flat-out get a bellyache. That's not to mention that when you're really thirsty after a long day, sometimes a beer in hand is just asking for trouble!

We've done you a solid by inventing this drink. After years and years of experiments involving more pints than we care to admit, one night we were tinkering in the shop and the idea struck. Instead of getting all Mel Gibsoned in a hurry, why not mellow out with a smooth blend of beer and 'buch!

1 cup refreshing light beer (we like Kelso Pilsner or a Belgian white ale such as Southampton's Double White)
1 cup kombucha
¼ lemon

Fill a pint glass first with Pilsner, then layer the kombucha on gently. Squeeze the lemon into the frothy drink, then drink the frothy drink. Darts, anyone? My money is on the 'buch drinker.

Kombucha Brooklyn Iced Tea

Serves 1

This remake of the classic liquor frenzy from Long Island takes all the same punch but adds a perfect balance of acidity and detoxicity. *Soooo* Brooklyn!

½ ounce vodka
½ ounce white tequila
½ ounce gin
½ ounce white rum
½ ounce triple sec
Ice
2 ounces kombucha

Mix all the liquors in a tall glass, add ice, and float the kombucha on top. Tip your hat to the best borough on Earth!

’Buchkey Sour

Serves 1

Whoever said a masterpiece is complicated was just lying. Whiskey, ’buch, and simple syrup—like all of the great triptychs, it’s as if they were made for each other. For the first-time kocktail drinker, make this simple work of art.

1½ ounces bourbon
2 tablespoons kombucha vinegar
½ ounce simple syrup (1:1 ratio of sugar to water by volume)
Ice
Lemon twist
Fresh cherry

Combine the bourbon, vinegar, and simple syrup in a cocktail shaker, add ice, and shake. Pour into a chilled cocktail glass and garnish with a lemon twist and cherry.

Dyke-otomy

Serves 1

Our good friend and Brooklyn neighbor lesbian party producer Sir Sabrina Haley came up with this one to represent the "work hard, work out hard, play hard" mentality of New York City.

In Sabrina's own words: "We like to party, but we still need to rock our next day and stay healthy to do it all. Since alcohol breaks down to sugar in the body, it's best to reduce overly sweet mixers and sugars in our cocktails. Agave is a healthier sweetener. This recipe also uses a crisp, clean silver tequila that acts as an energy enhancer as a result of the process in which it is distilled from the agave plant. Most liquors are depressants in comparison. The base is none other than healthy, detoxifying kombucha with all of its nutritional benefits."

6 ounces silver 100% agave tequila
2 ounces kombucha
Ice
3 tablespoons Cinnamon Agave Syrup
(see page 161)
½ lemon
Splash of seltzer
1 orange slice
Dash of cinnamon

1. Pour the tequila and kombucha into a cocktail shaker over ice. Add the Cinnamon Agave Syrup, squeeze in the juice from the lemon, and shake for about 20 seconds.

2. Strain the mixture over ice into a cocktail glass, top with a splash of seltzer, garnish with the orange slice, and sprinkle the cinnamon on top for added beauty and digestibility.

CINNAMON AGAVE SYRUP

1 cup agave nectar

1 cup hot water

1 tablespoon ground cinnamon

Combine the agave nectar and hot water and stir to dissolve. Stir in the cinnamon until completely dissolved. Let cool completely before using. Store in the refrigerator between uses for up to 1 month.

Kombucha Mint Julep

Serves 1

This drink is the work of the kombuchman's dad. A romantic soul creates a romantic kocktail. Drinking this will leave you pining for a southern porch on a late summer evening with some good company and perhaps a game of rook.

1 sprig fresh mint
2 ounces bourbon whiskey
4 ounces kombucha
1 teaspoon raw local honey dissolved in 1 tablespoon warm water
3 ice cubes
Orange peel
Strip of lemongrass

Muddle the mint leaves in a lowball glass. Gently shake the whiskey, kombucha, dissolved honey, and ice in a cocktail shaker and pour over the muddled mint. Spear the orange peel with the lemongrass strip for an aromatic garnish.

Cherry Kombucha Blossom

Serves 1

In late spring along the East Coast we are blessed with the most beautiful display of nature with the blooming of the cherry blossoms. This stunning and refreshing time inspires cherry blossom festivals and celebrations around the globe. Here in Brooklyn, we celebrate the cherry blossoms in numerous ways, but perhaps the most famous is the festival that the Brooklyn Botanic Garden throws. The first year KBBK was involved, we came up with an amazing pink and floral Cherry Blossom Kombucha. The blend was a hit, and around the KBBK labs, it was quickly kombined to make a beautiful kocktail.

2 ounces vodka
6 ounces cherry-flavored kombucha
Ice
Lemon peel
Splash of orange bitters

Shake the vodka and kombucha over ice. Rub the rim of a martini glass with the lemon peel. Strain the mixture into the martini glass, splash with the orange bitters, and toss the lemon peel into the center of the drink for a garnish.

'Buchtween the Sheets

Serves 1

1 ounce white rum

1 ounce brandy

1 ounce triple sec

1 ounce kombucha vinegar

Ice

Lemon twist

Shake all the liquid ingredients in a cocktail shaker with ice. Strain the mixture over ice into a chilled cocktail glass. Garnish with a twist of lemon.

Kombu Kollins

Serves 1

A classic cocktail turned kocktail, this drink will bring out the sneaky trickster in you. The Great Tom Collins Hoax of 1874 had merry people chasing a fictitious loose talker all over New York City. Our funky twist on this pre-Prohibition drink is yet another turn in the centuries-long search for the elusive mister.

¾ ounce kombucha vinegar
½ ounce simple syrup (1:1 ratio of sugar to water by volume)
2 ounces gin
Ice
5 ounces kombucha
Lemon twist
Fresh cherry

Shake the vinegar, simple syrup, and gin in a cocktail shaker over ice. Strain over ice in a tall glass, top off with the 'buch, and garnish with the lemon twist and cherry.

Warm Mulled Kombucha with Brandy

Serves 1

In the winter months we all need a warm and soothing beverage on our side table while we while away the hours knitting and playing poker. This kocktail will do just that. And for the kiddies, you can dish out some warm mulled kombucha before the brandy gets added.

3 ounces kombucha
3 ounces apple cider
1 teaspoon mulling spices in a tea ball (see sidebar on page 167)
¼ teaspoon vanilla extract
1½ ounces brandy (optional)

1. Combine all the ingredients except the brandy in a small saucepan and carefully heat at a low temperature, testing the temperature often with the tip of your finger. Once the liquid feels nice and warm to the touch, remove the pan from the heat source. If you have a thermometer handy, use it to make sure your 'buch doesn't creep above 110°F. This will preserve the raw lifefulness of your brew.

2. Remove the spice ball and serve to the kiddies. For an adult beverage, add the brandy. Sit in front of a fire, inhale the vapors, sip often, relax.

A nice mulling mixture is useful and spirit soothing throughout the fall and winter months. Warm it with kombucha, apple juice, pear juice, orange juice, or even red wine for an aromatic spin on normalcy.

1 tablespoon whole allspice seeds

1 tablespoon whole cloves

5 cinnamon sticks, crushed or chopped

Seeds from 6 cardamom pods

Zest of 1 orange

KOSMETICS AND KOMBEAUTY

Ever notice the strong connection between your skin and your gut? For instance, the kombuchman is a Nordic dude. Many Nords have an intolerance to dairy, and the condition presents itself as eczema. So the kombuchman eats dairy and his skin freaks out. What's up with that?

We talk a lot about skin being an important barrier, but we don't as often appreciate the gut's role in protecting us from the outside world. Think about it: You stick forkfuls of outside material into your body every day via the "food tube." The skin and GI tract are really one continuous sheet that separates you from the harsh microbial world, whose sole objective is to infiltrate, multiply, and take over!

The GI tract lining and skin are primarily made of the same cell type: epithelial cells. These cells communicate with each other chemically through the bloodstream, so when you eat something that makes your gut lining unhappy, your skin gets the message and also has a meltdown.

Kombucha nourishes and supports epithelial cells with acids, enzymes, antioxidants, vitamins, and pH-adjusting capabilities. It's obvious when you drink 'buch that your insides are being well taken care of, but have you noticed that kombucha drinkers have firmer, more gorgeous skin? Well, kombucha bathers have truly radiant skin! A few savvy skin care companies have picked up on this by creating kombucha cleansers, moisturizers, and serums. What follow are recipes to make your own fresh kombucha-based beauty products at home.

Kombucha Is Not Just for the Kitchen. Brew Some for Your Bathroom Too!

Kombucha Clay Mask

The fine texture and purity of calcium bentonite clay is truly luxurious. Use any 'buch home brew you have handy. Underfermented, overfermented, and perfectly drinkable 'buch—they all have their benefits. Sugar still present in underfermented 'buch is a natural skin brightener containing alpha-hydroxy acids that condition and moisturize the skin. Overfermented vinegar-style 'buch is a tonic and promotes blood circulation, reduces scaly or peeling conditions, and regulates the pH balance of skin. If your 'buch is in between, you get the best of both worlds. Mix in dried lavender, rosemary, rose, or elderflower, if you have them, for herbal and aromatherapeutic properties.

2 tablespoons calcium bentonite dried clay (available on the KBBK website)
2 tablespoons kombucha
½ teaspoon dried aromatherapeutic herbs (optional)
2 slices cucumber (optional)

1. Mix well all the ingredients except the cucumber in a ceramic or glass dish.

2. Apply a nice thick layer to your face and neck. Relax for 10 minutes or so with the cooling, soothing cucumber slices over your eyes. Listen to Tropicália. Dream of the healthy ocean air.

3. When the mixture starts to dry, you will feel your skin coming alive! Rinse the mask off with water, use a good toner, perhaps a kombucha-based one, and moisturize with your favorite lotion or all-natural oil. Voilà . . . you are gorgeous!

Kombucha Soak for the Sole

Feet. Some people deck them out with anklets, pedicures, and henna. Others neglect them, burying them in the damp darkness of socks and shoes, setting them free only to shower. Whichever foot personality type you are, it is essential to recognize the importance of feet! They are the structural foundation of the rest of your body. When your feet lose their strength and flexibility, the rest of your body cannot achieve whole balance and vitality.

Balance the forces from above with the forces from below with this Soak for the Sole. It will ward away foot fungus and soften feet while it brings you closer to your divine source.

2 tablespoons coarse sea salt
2 tablespoons extra-virgin olive oil
5 drops jasmine essential oil
Tennis ball
3 cups kombucha vinegar
1 sprig rosemary
10 drops balsam fir needle essential oil

1. In a small dish, mix the salt with the olive oil and jasmine essential oil. Set aside.

2. Place the tennis ball on the floor and stand next to it. Roll your right foot across the ball at a pretty good pace. Go from side to side and from front to

back. Ground your heel and let your toes roll back and forth over the ball. Awaken and release the tension in both of your feet this way, spending at least 2 minutes per foot.

3. Fill a foot-size soaking vessel with hot water. Make it as hot as you can stand. In my house, the tap water gets just hot enough. If yours isn't quite hot enough, heat some water on the stove. Gently add the vinegar, rosemary sprig, and balsam fir needle essential oil.

4. Place your vessel in an area where your feet can relax. Although it is nice to soak your feet while sitting, you can also try soaking your feet while lying on the floor on your back with your knees bent. This is a great way to level the playing field of your body and a good place to meditate. Balsam fir needle essential oil is mood and spirit elevating. Allow relaxation.

5. Dunk your feet and soak. Tickle your feet with the rosemary sprig, perhaps threading it between your toes. Remind your feet that they are loved. Feel the kombucha going to work on your skin. Let the microbes clean your feet like a suckerfish in an aquarium. After 10 minutes of soaking, treat your feet to a jasmine salt scrub. Pat your feet dry and gently massage the salt mixture onto your feet and calves. Don't forget to get between the toes!

6. No need to rinse the salt away; just plop your feet back into the soak and relax until your feet are content. Be careful when you take your feet out. They will likely be slippery. Pat them dry gently, appreciate how smooth and supple your feet can be, and carry on with your barefoot day!

Kombucha Locks

Wellness of the mind, body . . . and hair! Make a statement of holistic health with luxurious and vibrant hair. Being healthy means paying respect to all of the states of our being. Humans love to look good and should relish in taking care of their visual assets, as we are all visual creatures! Don't call it vanity. Let your beauty rise to the surface. Be sublime! What is sexier than hair that screams health?

> Like skin cells, hair cells are also epithelial cells and respond well when they're nourished by the same probiotics, prebiotics, vitamins, and enzymes that the gut loves!

Half a gallon of kombucha culture
1 cup kombucha
Juice of ¼ lemon
½ cup raw rolled oats
2 drops vitamin E oil
2 drops lavender essential oil

1. Combine all the ingredients in a blender and blend until smooth.

2. Wash your hair in the shower with a natural shampoo and rinse.

3. Work a dollop of the kombucha hair tonic into your hair and scalp. Be sure to pay extra attention to your scalp and get a good massage out of it! Let stand for 5 minutes while you take the time to sit and rub your feet.

4. Rinse well. Once it is dry, look in the mirror, toss your hair around, and say to yourself, "Damn, you look good" . . . because you do.

KomBody Mist

Body sprays are nice for cooling down and freshening up in the summer. In the winter, a good misting humidifies the skin cells and helps keep your body moisture intact. Create your own signature nourishing body mist by blending essential oils that you love and that are seasonally appropriate. Two good blends for winter are 2 parts frankincense: 1 part grapefruit and 1 part orange: 1 part peppermint: 1 part rosemary. For summer, try 10 parts sandalwood: 1 part jasmine and 1 part lavender: 1 part chamomile.

30 to 40 drops essential oils
1 tablespoon witch hazel
½ cup kombucha
2 cups filtered water
Coffee filter
Small spray bottle, preferably glass

1. Blend the essential oils and witch hazel in a small glass bowl. Mix thoroughly and allow to stand, covered, for 30 minutes, allowing the scents to blend and the alcohol from the witch hazel to evaporate.

2. Add the kombucha and water. Mix well, cover, and allow to stand for 10 more minutes.

3. Pass the mixture through a coffee filter, then carefully pour into your spray bottle. Shake well before using. Refresh your supply every week.

Whole KomBody Detox Bath

A whole body detox should be done with the changing of every season. Some people fast, some people schvitz, some people pray. As 'buch brigadiers, we do all those things and more . . . and we do it from the inside out.

Whatever your program, a soothing, rejuvenating, and nourishing Whole Kom-Body Detox Bath will up the ante. Prepare for a great season anytime with this opulent soak.

2 cups kombucha or kombucha vinegar
A bathtub full of comfortably hot water
2 tablespoons fresh ginger juice
30 minutes to spare

1. Add the kombucha to the bathwater, then add the ginger juice. Add you! And soak for as long as you wish, but for at least 15 minutes. Don't bring your cell phone or your iPad. Perhaps dial in some chanting monks on your stereo and be sure to take in the mental and spiritual cleanse along with the body detox.

KomBlanket Time Machine

This daily practice will bring blood to the skin where kombucha can exchange with it the secrets to a long and gorgeous life. It's like a personal spa just for your face that is subtle enough for daily use. If another area of your body craves this kind of attention, let her have it. Desire is usually there for a reason. Varicose veins or sore muscles will sigh with relief with this treatment as well.

The KomBlanket Time Machine should be added to your skin care routine either upon awakening or just before turning in at night. Expect to notice the difference immediately, and remember to smile coyly when your crush can't keep his/her eyes off of you.

Washcloth
Comfortably hot running water
1 teaspoon kombucha or kombucha vinegar

Soak your washcloth in the running water. Ring it just enough so water is not pouring out of it. Sprinkle the kombucha over the washcloth. Press the warm KomBlanket Time Machine over your entire face for 30 seconds or longer. Be sure to breathe in the secrets as well. Rinse. Repeat if desired, and finish with a natural toner and moisturizer.

EPILOGUE

As kombucha continues to penetrate the Occidental psyche, new and more technically marvelous uses for kombucha and the SCOBY itself are being developed. Research currently being done on the health benefits of drinking (and eating) the fantastic tea-based elixir will shed more light on the increase in life force kombucha drinkers have known for generations. Perhaps new studies will highlight the specific therapeutic qualities of kombucha too. We hope so, and we encourage anyone who is interested in a rigorous study involving kombucha to contact us for samples and consultation.

In addition to kombucha's role as a healthy beverage, food, and beauty product, innovators around the globe have been thinking outside the jar. The future of kombucha will include industrial applications in environmentally and socially responsible products. From children's clothing in fabrics made from kombucha SCOBY cellulose grown in small sustainable fabric farms to using the heat of the exothermic kombucha fermentation to gently warm sprouting seeds, art supplies, or compost piles, the benefits of kombucha are still being discovered. Innovate! Design the new millennium to your likeness where nourishing things such as kombucha can provide us with the joys of modern living!

As 'buch brigadiers, we hope you continue to grow your knowledge and spread it around. Like the ancient ferment being passed down from generation to generation to finally land in your cup, continue to pay it forward and share the health and wealth of possibilities of kombucha!

KOMBUCHTIONARY

Bacteria Microscopic single-celled organisms that don't have an organized nucleus or organelles. Bacteria are everywhere! There are ten times as many bacterial cells in your gut as there are human cells in your body. The vast majority are harmless to humans, many are beneficial, and some can cause disease. Bacteria are known for food spoilage, but most ferments are the result of this same bacterial activity done in a way that enhances the flavor, nutrition, and digestibility of the substrate (see *Substrate*).

'Buch The delicious drink called kombucha.

'Buch brigade A group of kombrewers led by the kombuchman, who brew and share kombucha with the people.

'Buch Kamp A place of learning and growth where you will receive all the knowledge you need to make the best damn kombucha in town.

Home brew Any brew made in the comfort of your home. *Mmmm,* tasty.

Kocktail An alcoholic drink consisting of a spirit or several spirits mixed with kombucha and perhaps some other delicious ingredients. We like to think of kocktails as reverse toxmosis (see *Reverse toxmosis*).

Kombang What happens when something involving kombucha gives you an experience worth punctuating!

Kombo An awesome mix involving kombucha. Whether it's a mad beat on the tables or a sweet blend of flavorings, a kombo aims to please.

Kombrewer Someone who brews kombucha with skill and finesse.

Kombucha A nutritionally alive drink that is full of compounds that detoxify, energize, support your immune system, nourish your digestive system and skin, prevent disease, and elevate your mood. Kombucha is also the subject of an awesome book by Eric and Jessica Childs. Have you heard of it?

Kombucha vinegar Kombucha that is too acidic for your taste, usually because of fermentation that goes on for too long or at temperatures that are too high. Useful for flavoring in kooking (see *Kooking*).

Kombuchasseur Someone with a palate attuned to the complexity of 'buch. One who is easily able to discern the quality and strength of a particular kombucha brew.

Kombuchman aka Eric Childs The man who is one with 'buch and set to share it with the world.

Komplow To move effortlessly and powerfully (through something) by the grace of kombucha. Used in a sentence: "You komplowed through that solo!" or "Your beauty komplows me, my princess."

Kooking Using kombucha to enliven your food with flavor, nutrition, and all-around awesomeness.

Mise en place French for "put in place." When brewing kombucha, *mise en place* means having everything you will need clean and out on the counter before you begin.

Nute A nutrient-dense solution, or substrate, that microbes transform during fermentation. For kombucha, the nute is sweetened tea. In beer making, the nute is called a wort. (see *Substrate*).

Reverse toxmosis Doing the detoxing and the toxifying at the same time. Commonly used to describe the process you undergo while drinking a kocktail (see *Kocktail*).

Substrate The environment in which or on which an organism lives.

Supreme kombulord Master of ceremonies at a party where reverse toxmosis is taking place (see *Reverse toxmosis*).

Yeasts Microscopic (mostly) single-celled organisms that have organized nuclei and organelles. Although yeasts aren't quite as ubiquitous as bacteria, they are all over the place, especially on the outside of sweet fruits and carbohydrate-rich grains. Yeasts generally are not a threat to humans, except in situations where the immune system is compromised. In those cases, a few varieties of yeast will take the opportunity to proliferate. Yeast is most commonly known for its role in beer, wine, and bread making, where it converts sugars into alcohols.

ACKNOWLEDGMENTS

Jessica and Eric would like to thank, in no particular order:

Gunther Frank and Michael R. Roussin for their passion and research on the subject of kombucha. Sam Dibble and Will Savitri for sharing our dream and holding our commercial production together. Jason Ashlock, who deserves spirited thanks for hitting it out of the ballpark—home run! Marisa Vigilante for hopping on board with a sharp, clean knife to cut away the unnecessary parts. Everyone who has worked for KBBK: without you, this book would not exist. Gary, Anita, Pat, and Bob (our parents) not only for your support but also for your encouragement and insight through this wild and crazy journey. Josie, Kees, Kasia, David, Aaron, and Ellen (our siblings) for being our dear friends in addition to family. Sandor Katz for showcasing fermentation so enthusiastically and for unknowingly introducing us to each other. Our tiny mansie, Rider Moselle Childs, for learning to pronounce "kombucha" flawlessly before he could say much else.

And last, we would like to thank each other. With you, I can be everything.

INDEX

Page numbers in italics indicate illustrations.